DEFACING POWER

DEFACING POWER

The Aesthetics of Insecurity in Global Politics

BRENT J. STEELE

The University of Michigan Press • *Ann Arbor*

First paperback edition 2012
Copyright © by the University of Michigan 2010
All rights reserved

Published in the United States of America by
The University of Michigan Press
Manufactured in the United States of America
⊗ Printed on acid-free paper

2015   2014   2013   2012   5   4   3   2

A CIP catalog record for this book is available from the British Library.

Library of Congress Cataloging-in-Publication Data

Steele, Brent J.
    Defacing power : the aesthetics of insecurity in global politics /
Brent J. Steele.
        p.   cm.
    Includes bibliographical references and index.
    ISBN 978-0-472-11732-1 (cloth : alk. paper) —
    ISBN 978-0-472-02280-9 (e-book)
    1. Security, International.   2. Power (Social sciences)
    I. Title.

    JZ5588.S742   2010
    355'.033—dc22   2010014121

ISBN 978-0-472-03496-3 (pbk. : alk. paper)

For Annabelle Kathleen and Joseph Eldon,
the genuine beauty of my life

Preface and Acknowledgments

The idea to pursue the intersection of aesthetics, power, and insecurity in global politics came to me not in an inspirational moment or even over a particularly beneficial conference or seminar. Rather, the idea for this book came over the course of several years and through observations in a variety of environments. These included the particular attitudes of individuals, groups, and various societies where appearance was considered an ample replacement for substance; the proclamations that "will" shows strength while backing down from a confrontation demonstrates weakness; and, of course, the cocky and bombastic veneer of particular world leaders who were in every other sense struggling to hold their countries together. This led to a hunch—that the confidence of the powerful is a facade, that what we often take to be strength belies a particular weakness, and (to paraphrase General Patton) that the security found in this aesthetic power is fleeting. Someday, I figured, I would investigate the facade of aesthetic power as it is manifested in global politics. This book is about the defacing of that power, what occurs when the *appearance* of strength is challenged.

Individuals do things to make themselves more beautiful or "presentable" to others (dress nicely, put on makeup, have good hygiene). By doing so, they feel more in control of who they are and what they are doing. But that sensation can be quite fragile. This book is about how powerful actors, including nation-states, do this as well—self-create in order to attain a sense of control, a *sense* of security. By challenging that self-creation, what I term *counterpower* creates quick moments of insecurity, forcing powerful actors to self-create yet again. A turn to aesthetics, where sensations are at the forefront, is potentially useful here, for our understanding of security, the activity of self-creation, and reactions to insecure moments in international politics.

Writing a book with the word *aesthetics* in the subtitle is risky even in today's more open and pluralist field of International Relations, and I surely could not have done so without the assistance, support, and friendships of many people. Several individuals looked over the prospectus and provided very helpful feedback, including Mlada Bukovansky, Ronald Krebs, and Francois Debrix. I also had the honor of receiving Francois's insights at the 2008 convention of the International Studies Association in San Francisco, where he served as a discussant and issued comments over a paper I presented on parrhesia that eventually migrated into chapter 4. In the summer of 2008, I benefited from a very professionally inspirational meeting at the "Thinking With(out) Borders" conference on international political theory, held at the University of St. Andrews. While there, I received helpful feedback from a number of individuals, including especially Tony Lang, who, along with Ward Thomas, gave some insightful advice while we were looking for Tony's errant golf shots in the gorse one afternoon. Tony also issued comments over chapters as the project evolved. He has served as a very important mentor and friend to me over the years, and I am fortunate to have met such a kind individual, not to mention world-class scholar, as Tony. Kate Schick and Scott Nelson supplied very helpful suggestions during my paper presentation at St. Andrews. I was also quite lucky to spend some time in Scotland with Nick Onuf, a scholar whose work has transformed the field of IR theory. Nick provided enormously helpful critical comments on several portions of the work.

Several other individuals, Stefano Guzzini, Benjamin Herborth, Patrick Thaddeus Jackson, Daniel Nexon, and Ole Wæver, through their work and also the discussions I have had with them, provided me a ton of food for thought regarding power and aesthetics. I have a group of friends who sent me detailed comments on several chapters. Jon Acuff and Jon D. Carlson not only imparted erudite, detailed, and (as always with those two) entertaining observations on draft chapters but also assigned portions to their students. I thank both of them and their students for this thoughtful assistance. Jack Amoureux, Jeremy Youde, and David McCourt supplied critical but incredibly supportive insights on portions of the work. Andrew Hom, a former grad student of mine who is now on to bigger and better things, somehow found the time to look over an endless array of chapter drafts that I sent him and issued fantastic and supportive comments. At the University of Kansas, I have had the luxury to be surrounded by a very impressive array of scholars (and friends) who have lent collegial and professional support. Most especially, I would like to thank Hannah Britton, Alesha Doan, Don Haider-Markel, and Julie Kaarbo. I thank Ben Lampe for formatting assistance.

The staff of the University of Michigan Press has been outstanding. I have no idea when Melody Herr finds time to sleep, but she is the most efficient and impressive editor I have met in this business of academic publishing. Melody procured several brilliant blind reviewers for this project at several stages of its development, issued her own helpful and grounded insights on several chapters, and was a critical but supportive voice while I kept plunking away at draft chapters, response letters, and marketing materials. Thanks to Melody for taking a chance on this work and to Scott Griffith for his very patient and efficient help in seeing it to production. I would also like to thank the three anonymous reviewers who provided enormously helpful feedback on this project.

Portions of chapter 2 are rewritten from my article "Making Words Matter: The Asian Tsunami, Darfur, and 'Reflexive Discourse' in International Politics," published in *International Studies Quarterly* 51, no. 4 (December 2007): 901–25. I thank Wiley-Blackwell Press for permission to republish. Portions of chapters 1 and 4 are rewritten from my article "'Ideals That Were Never Really in Our Possession': Torture, Honor, and U.S. Identity," published in *International Relations* 22, no. 2 (2008): 243–61. I thank Sage Press for permission to republish.

I have benefited from a very supportive network of close friends and family, including my parents Ted and Barb Steele. The many golf outings I have had with my friend Matt Pogemiller have proven to be welcome diversions from academic life, and I thank him for tolerating the continued deterioration of my game. My brother, Kyle Steele, his wife, Lisa, and my nephews, Brenan and Kaleb, are a constant source of love, entertainment, and encouragement. My grandfather Eldon Akers continues to inspire me. A lot of what I am today is due to the unconditional love and support he provided me for over three decades.

Finally and most important, my wife, Mindy, once again proved to be the rock of our family during the many late nights (and early mornings) I spent at my office on writing binges. She was and always has been a steady source of support and encouragement. My children, Annabelle Kathleen and Joseph Eldon, enrich my life and impact it in ways they will never know. Both coming home to Belle during the last weeks of drafting the first full manuscript and the many days with Joe during the months I spent revising provided fuel for an often exhausted soul. Their beauty, laughter, and love have transformed me for the better. It is to them that I dedicate this book.

Contents

Introduction

The field of International Relations often regards powerful nation-states as rather stagnant entities, robust in their ability to maintain authority and relative control over time and space. This is not without good reason, of course, as these actors possess an enormous ability to influence others in international politics and, more pressingly, deploy a substantial amount of organized violence. This static assumption is usually understood with reference to their material or strategic position in the global order, and their decline is brought about by challenges from other actors, through free riding or through force (Kennedy 1987; Gilpin 1981). To this end, the fields of international relations and international history have helpfully established the manner in which international actors can challenge those in authority through "balancing" behavior (Spykman 1942; Waltz 1979; Walt 1987). These challenges do, however, take substantial periods of time—decades, even centuries—to emerge. Further, if the material and strategic capabilities of states constitute the "power" that serves to deter would-be challengers, why do the most powerful states or ideologies appear to be so routinely challenged? Why do these actors perceive their supremacy to be under constant challenge? Why do they appear to be so caught up with their self-image? How does power in this way facilitate its own challenges?

There is another type of insecurity that the powerful in international politics face—an aesthetic insecurity over appearances. This insecurity is more frequent, localized, and public. These instances where centralized bodies of power, including nation-states, react to aesthetically problematic representations have been underanalyzed by IR scholars. While they do not lead to substantial shifts in global distributions of power, this book advances the claim that these moments of aesthetic insecurity are still worthy of study, if we understand these reactions to be a logical tendency of power itself. In doing so, the book posits the possibility that processes other than state-based balancing can manipulate centralized bodies of power, stimulating them into new modes of behavior. Because these bodies brand themselves aesthetically and because their citizens or group members draw intrinsic psychological, rhythmic, and imaginative satisfaction from such constructions, they can be manipulated through what is advanced in this book as *counterpower,* moments when those aesthetic visions are challenged or ruptured. Though examples are drawn from the aesthetically created Self of U.S. power for illustration, the broader implication of the argument is that all bodies of centralized power can be countered aesthetically or ontologically insecuritized through counterpower. In a nutshell, I argue in this book that centralized power, including but not limited to national power, recognizes its own ability to self-create; that the aesthetic construction of this ontology is engaged by collective bodies of citizens; and that when such a construction is de-aestheticized, these bodies react in quick and sometimes problematically violent ways.

To position this argument, let me clarify a couple of preliminary issues. First, what is meant here by the "Self" of power and nations? The term may seem a bit jarring to readers unfamiliar with the debates over the notion of a "Self" of groups or nation-states. My previous work (Steele 2008a) and the debates that in part inspired it[1] grappled with the problems and benefits of ascribing the human Self and its biological, psychological, and sociological drives to corporate actors such as the nation-state. These meditations have hardly resolved this debate, and the current book will do no better in satisfying those skeptics who find it analytically problematic to ascribe individual Selves to the nation-state or other corporate bodies. Nevertheless, we can recognize colloquially how we talk everyday about corporate actors as if they are people, as in "France *recognized* the Armenian genocide," "Did you see what Israel is *doing* in Gaza?" or "Apple *introduced* a new iPod today!"[2] This does not assume that they are "real" people by any means, but as corporate actors, they engage in *practices* that are real and observable, and these practices occur in what one scholar terms "contested zones of ongoing debate [rather] than physical spaces" (Jackson 2004: 285). Sometimes we go a

---

1. See special issue, *Review of International Studies* 30, no. 2 (2004).
2. Alexander Wendt (2005) has made this point more forcefully than I ever could, in his response to Peter Lomas's (2005) critique of state personhood.

bit further, ascribing a motive to these actors, again as if they are individuals, as in "France did this because it has a large Armenian population," "Israel is trying to get Hamas to stop launching rockets" (or "Israel is trying to demonstrate force to Iran"), or "Apple is trying to corner the market."

In this book, I continue to maintain, as others have (Lang 2002), that because state agents "narrate" about the nation-state, they create potential Selves that that nation-state seeks to realize through its policies. Yet in contrast to my previous work, the current investigation, rather than engaging in the methodologically problematic posture of ascribing a human motive or intent to a corporate Self, identifies a particular *practice or activity* of individual self-creation at the level of centralized bodies of power, which includes not only nation-states but other organizations as well. Instead of engaging in the more difficult task of ascribing motives to a "Self" of nations or other groups relevant to international politics, then, I seek to understand how the activity of aesthetic self-creation, if it obtains in international politics, makes possible certain processes and outcomes that require explication.

A second preliminary issue is to tackle what exactly is meant by an *aesthetic* practice. I discuss why this move is appropriate in international relations later in this introduction, but for now, let us start with how the aesthetic functions for the individual. We use aesthetics ourselves to be more attractive to others through style—we pay attention to forms of hygiene; we exercise; we self-monitor the way we talk, dress, and "act." We may engage in these practices or activities because they open up strategic possibilities for us—we want others to "like" us, we might want someone to hire us, we want to get a raise, we want others to be happy, and so on. But these practices also generate a sense of self-worth—in creating ourselves, our perceived Selves create us. This fashioning of ourselves as aesthetic beings is something we can admire on our own *in addition to* the function it serves in forcing others to see us differently. It is this practice of self-creation as it works back upon the individual that I ascribe to bodies of power. The insecurity arises because the purchase of aesthetics is to make objects *appear differently* than they otherwise were. The term we use for cosmetics—*makeup*—captures this quite nicely. When we are stripped of these aesthetics, we are vulnerable—not only to others, but also to how we see ourselves. This occurs for most of us first thing in the morning, so we avoid the mirror until the moment when we can begin refashioning ourselves for the day ahead.

Individuals, when aggregated into groups, also seek out this process of aesthetic creation. Corporations, for instance, stay edgy by updating their logos to keep their "brands" fresh. They adjust their marketing slogans and acquire different celebrity spokespeople, for the nature of celebrity itself is momentary. The definition of *aesthetics* as a noun is suitable here: "a philosophical theory or idea of what is aesthetically valid at *a given time and place.*"[3] Here, aesthetics refer to a contextualized study of the "moment" in which this creation takes place. The activity of creation is never fixed and is subject to constant updating. The point is not that a corporation is a living, breathing "Self" like an individual human being is but that it engages in a similar practice of aesthetic self-creation. The corporation creates a "brand" that is more than just the products it provides to a consumer.

When we get to international politics—where organized violence is still considered to be a legitimate practice—we can identify particularly unique manifestations of this practice of self-creation and the mode in which aesthetics intersects with power. Nations may acquire power for a variety of purposes, but when it is used and exercised, the full contours of the national Self become apparent (Niebuhr 1932: 96). Yet when we analyze aesthetic construction in corporate actors—and all that construction makes possible—we are focusing on how the individuals who form such an organization participate in the creation of the *corporate* self. More than the material basis of power, the aesthetics of power is made possible by the psychological and emotional connections that humans have toward a corporate body of power.

The aesthetic functions here as it does for the individual, it creates the Self *other than it may be,* a construction perpetually contingent on the moment and always subject to a process that might reveal it to be something else. This does not mean that there is a "core" Self of national or group power that can ever be realized—in fact, the urgency behind perpetual creation indicates that there may never be a core or timeless Self that can be accessed. In other words, there is never a once and *true* Self that emerges in this process, because self-creation is an activity that engages a Self that is always ambiguous. Aesthetics attend to this constant flux of the Self and the insecurity that emerges from its centrifugal tendencies. In an age of instantaneous and continuous global and publicized communication, there is no privacy for re-creation. Instead, in con-

3. http://dictionary.reference.com/search?q=aesthetic, emphasis added.

trast to the individual who does so in their own home before the day begins, the Self of such organized bodies re-creates out in the open.

This understanding of the intersection between aesthetics and power assumes, then, that any power that appears to us in the form of "self-certainty, self-aggrandizement, and crusad[ing]," as one vibrant account describes the contemporary U.S. Self (Tjalve 2008: 2), is a power that poses as confident. When power manufactures the aesthetic, it in turn belies a sense of security and certainty. Just like the process of aesthetic self-creation the individual engages in, the aesthetics of power is manifested not just in images but also in styles of discourse.

These aesthetics can be challenged through what I designate as *counterpower*, those moments when the aesthetic creation of the Self is challenged, forcing power to re-act rather forcefully and quickly. Return again to the individual—when our aesthetic integrity is disrupted, we sense an intense need to re-act, to *move* away from that previous depiction of our Self. If there were such a thing for individuals as a true Self to lock into, we would never feel the need to refashion who we are; we would just "be ourselves." But of course this is not possible, and if this process of aesthetic insecurity operates in international politics, then regardless of how certain or assured powerful nation-states seem to us, no matter what type of facade—or "face"—they advance, they do so to cover up a particular form of vulnerability that can be manipulated. In short, this book investigates how power uses aesthetics to "make up" for an inherent insecurity—an insecurity that can only be observed when such aesthetics are deconstructed.

how aesthetics "make-up" for own inherent insecurity

The Purchase of Aesthetics and Power in International Relations

Why has the field of International Relations not fully explored this possibility? The lacuna is due in part to the field's more pressing preoccupation in recent years with power balancing. Following the Cold War, while many theorists predicted a challenge to U.S. hegemony, scholars struggled over the manner in which such a challenge would materialize. This concern with systemic power consumed much scholarship, even though, as this book suggests, aesthetic power was operating—and being insecuritized—in our plain view the whole time.

Take, for instance, the debates that occurred in the 1990s over the preponderance of American power and the manner in which the international community would come to terms with this imbalance. There were, generally, three camps that industriously confronted the issue. Neorealists such as John Mearsheimer (1990) saw an unstable international system that would attempt to balance against U.S. preponderance, and Kenneth Waltz (1993) wrote about an "emerging structure of international politics" after the Cold War thawed, drafting several scenarios where various blocs would emerge to counterbalance the United States. Christopher Layne was a bit more precise, asserting the unipolar moment to be "just that, a geopolitical interlude that will give way to multipolarity by 2000–2010" (1993:7).

A second group, also realists but ones of the neoclassical variety, saw a stability and durability of the unipolar world. William Wohlforth (1999) was one of several scholars who made this argument (see also Mastanduno 1997), pointing to the historically unique amount of asymmetric power that the United States possessed compared to other hegemonic examples and asserting that emerging powers would more likely trigger regional power-balancing dynamics before they would attain the status of a systemic rival. Thus, U.S. hegemony would remain the order of the day. A third group agreed with the second about the durability of U.S. hegemony but took a different approach to explain why this was so. Their answer was that U.S. hegemony was a benign presence—the character of U.S. hegemony as liberal, open, and institutional made it less threatening to other potential rivals (Ikenberry 1999). This reasoning was reinforced by the intrinsic nature of U.S. politics—a legitimate or benign democratic identity that helped cultivate and also allow for a transatlantic "security community" (Risse 1996; Barnett and Adler 1998).

Alas, after the 1990s ended, no balancing had occurred. The iconic publication that represented the scholarly "age of impatience" with this result was G. John Ikenberry's *America Unrivaled* (2002). Ikenberry introduced his volume's purpose as seeking to answer the question, "why, despite the widening power gulf between the United States and other major states, has a counterbalancing reaction not yet taken place?" (2002: 3). Literature emerging over recent years takes U.S. hegemony as a social fact but recasts it, even reconceptualizes it altogether, in terms of imperial structures (see Nexon and Wright 2007). This has come coupled with a set of studies that have reviewed how empires end, investigating where and when they become vulnerable (Spruyt 2005; Cooley 2005). The interest in balancing or the lack thereof continues to this day, as Randall Schweller's theory of underbalancing in *Unanswered Threats* attests. Schweller's study revealed that such underbalancing was more common than previously thought and could be explained by reference to domestic politics.[4]

Despite the seeming lack of an emerging "balance" to the United States at the systemic level, these debates have nevertheless been productive in delineating not only the distribution and flow of power at the international level but also how the meaning of such power can be transformed in international politics through institutions and discourse, as well as the manner in which *legitimacy* influences power's maintenance.[5] Yet centralized bodies of power respond to more than overall systemic distributions. If we shift our focus from the expansive terrain of empires and systemic power distributions to the ambiguous space of the aesthetic Self, we can find more observable phenomena that might prove just as worthy of study. If we shift our temporal lens from the decay or erosion of power in the decades and years toward days or moments, we can observe important vulnerabilities. To do this, we must appropriate the problem of power as a form of a transgression, a formation by the Self of its own limits, limits that are never realized until they are crossed.

I thus propose to add one further kernel of thought for theorists concerned about power and the vulnerabilities that it produces: when power is internalized within or melded to the Self of a centralized body (whether that is an individual, a group or organization, a nation-state or a hegemon, or even an informal empire), this internalization produces its own logic of insecurity that cannot be fully captured by external sources of threat. This insecurity is self-generating, bringing to bear a seemingly imperceptible force that nevertheless is far more prevalent than international relations scholars have acknowledged. Since at least John Herz's (1950) seminal essay on the topic, which has been recently thoroughly updated and extended by Ken Booth and Nicholas Wheeler (2008), the self-defeating logic of seeking infinite security writ large has been a central "dilemma" studied in international relations. Even more recent work has engaged the problematic but nevertheless real emotional processes (such as fear) that are embedded in this dilemma and in politics more generally (see Robin 2004). However, the internal vulner-

---

4. See also the continued interest of some scholars, such as Brooks and Wohlforth (2005), in this puzzle.

5. "The legitimacy of imperial rule matters a great deal for its persistence. As a form of hierarchical organization, empires need to convince significant actors that the benefits of continued imperial rule outweigh the costs of domination. In doing so, they make resistance less likely and thereby secure continued imperial control" (Nexon and Wright 2007: 264).

abilities that arise when the Self of a nation-state becomes aesthetically enamored with power have not been fully explored. In contrast to James C. Scott's argument about the "weapons of the weak" (1985), then, this book is about the "weakness of projected strength," or, more precisely, the vulnerabilities that arise from the aesthetic certitudes that color the image of power. Whereas in Scott's account, the resistance is localized, subterranean, and calculated, counterpower is publicized, in-the-open, explicit, and spontaneous.

Aesthetics, Power, and IR

What explains the move to *aesthetics* and power? The relationship between aesthetics and international relations became an increasingly vibrant prospective field for analysis, so that by the beginning of this decade, an article by Roland Bleiker (2001) could center on the "aesthetic turn in international political theory." Yet five years later, in another special issue of the critical IR theory journal *Millennium*, Gerald Holden noted that the field of aesthetic IR, while moving from critical IR theory outlets toward "a wider range of journals" (2006: 795), had also "been in existence for the best part of ten years, and those now joining the debate are not venturing onto *terra incognita*" (802). One of the problems, in Holden's view, was the work conducted by "literary" IR scholars who "used" literature to "bolster normative views they already hold about world politics" (794). Among Holden's many observations was how aesthetic "IRists" had tended to act as if they were engaging aesthetic subjects marginalized by mainstream IR theorists. On the contrary, Holden suggests, "many of the classical realists which had been targeted critically by aesthetic IRists in fact contained aesthetic expositions in their theories, which could be examined in a complementary fashion in further investigations" (802–3). Whether we accept Holden's view that aesthetic IR theory up until 2006 (at least) has been largely "preaching to the choir," my exposition of aesthetics and power (inevitably, in some respects) incorporates insights from several classical realist scholars, including Hans Morgenthau, George Liska, Arnold Wolfers, John Herz, and Reinhold Niebuhr.[6]

In addition to probing these theoretical intersections, the move to

---

6. Considering the role that prestige has played in classical realist formulations of power (see Markey 1999; Löwenheim 2003: 25–27), the use of these theorists to build an aesthetic understanding of power is hardly surprising.

aesthetics can be justified by reference to three further functions or purposes. First, a social scientific set of functions are served in that while the aesthetic is abstract and ambiguous in formation (necessarily so, as I point out), it nonetheless has an impact on the security practices of nation-states and on all movements of power in a broader sense. This book is written within the general tradition of critical security studies, and it draws on some giants of critical philosophy and social theory, but it does so in order to uncover a mechanism (aesthetic insecurity) that to date has not been comprehensively advanced. More broadly, while this book may be considered a critique of existing accounts in IR theory that posit how power is challenged, it is not a deconstruction of those accounts. I do not argue, in some fantastical way, that the material and strategic elements of violent power do not exist "in reality," nor is it my point to dispense with those accounts. Rather, this book argues that alongside these accounts, many of which are discussed in chapters 1 and 5 of the book, should be added an aesthetic account of insecurity.

These aesthetics are hardly purely "cosmetic," however. The idealized beauty of a centralized body of power, when compromised, can produce intense emotional and traumatic ruptures—ruptures that powerful actors seek to rectify by reacting in sometimes violent ways. This aesthetic layer is more than just the "style" or strut of power, then; it is, rather, the springboard that allows power to deploy its material and strategic resources in global politics. So while my study is critical in its reformulation of certain concepts used in IR theory (such as power and security), it is conventional in that it seeks to uncover what has become a largely unproblematized and common pattern of power's operation. In short, there is most likely an intuitive sense, by practitioners and theorists alike, that aesthetics intersect with power at all levels of human interaction. This study attempts to expound this intuition with careful analysis. There is an analytical interest in studying the aesthetic of the Self on its own accord—the aesthetic as an end ideal. At some point in time, certain nation-states accustomed to crafting their power to meet security threats shifted toward a more basic sensual celebration of the style and physique necessary for and found within *movement*. This leads to the romanticizing of the Self and thus to an idealized Self with much built-in vulnerability. This book is thus an investigation into the way in which the aesthetics of power lead to insecurity. It is a book that seeks to identify an aesthetics of insecurity.

If there is a social scientific purpose served in constructing the aesthetic layer of power, there is also a similar purpose served in understanding the processes or forms through which such power is challenged—what I term *counterpower*. This is also a "real" phenomenon in that we can recognize those moments when the aesthetic Self or subjectivity of power is ruptured or even unfastened. Thus, this book seeks to conceptualize and demarcate three particular categories of counterpower, termed hereafter as *reflexive and flattery discourse, parrhesia,* and *self-interrogative imaging.* To this end, I draw on a variety of philosophers (such as the late work of Michel Foucault) and, in a more limited sense, groups of social theorists, to articulate the basis for aesthetic power and counterpower. To understand counterpower, this book complements and critiques existing analyses of the resistance to power.

The second class of purchase for this book is derived from recent events and scholarly debates regarding the seemingly sophisticated and reasonable relationship between democratic publics and war. I reference here the research statistically suggesting that while they may be more likely to win wars (Reiter and Stam 2002; Bennett and Stam 1998), democracies are inherently vulnerable to protracted and drawn-out wars because the public, which presumably "calls the shots" in a democracy, removes its support (Gartner 1997). Further, "while autocrats can repress this dissent, democrats are forced to respond to it or face being replaced" (Filson and Werner 2004: 298). Others have pointed out how the free press of Western democracies allows hostile groups such as al-Qaeda an effective medium through which to manipulate public opinion. One scholar titles this a "virtual war," one where the public opinion of democracies can be manipulated and undermined, while insurgent enemies "have time on their hands and they will use this time against the West in the true tradition of Mao's protracted warfare" (M. J. Williams 2007: 274). While the purpose of these various arguments is quite different, of course, neoconservative writers have made the case (but in more ideological tones) that internal dissent makes a democracy inherently weak and that terrorists not only exploit this but know this to be the case and thus organize their strategies around this central assumption of "democratic weakness."[7] Yet this investigation compels us to equally notice how democracies, because their citizens perceive an ability to help in the fashioning of the national Selves, may also be unique in their vitalist tendency to overreact when the national Self appears to be in jeopardy. Such a primed posture makes these democracies vulnerable a priori, be-

---

7. A good example is James Arlandson, "The Logic of Weakness," *American Thinker,* June 2004, http://www.americanthinker.com/2004/06/the_logic_of_weakness.html.

fore a threat even emerges. This book helps us reformulate how overreactions are fueled not only by the material challenge al-Qaeda presents to U.S. sovereignty (in terms of its ability to use violence within the latter's borders) but also by its ability to manipulate U.S. power's aesthetic facade. The vulnerability of vitalism is addressed in chapter 1, and the counterpower possibilities of al-Qaeda are examined more directly in chapter 3.

The third purpose of this book is normative. Attached to an understanding of the vulnerabilities of bravado, strut, toughness, and physique is the side effect that those who seek security for national communities through such an ethos are instead putting those communities in a problematic position. Such emphasis brings little of the security it is at least *stated* to offer. The reader of the following pages will quickly notice that most of my empirical illustrations come from the U.S. foreign policies of the George W. Bush administration, yet I do not see aesthetic moves beginning (or ending) with that administration. This normative purpose requires us instead to "read ahead," as such aesthetics have not magically dissolved with the exiting of George W. Bush from the American political landscape. Spatially, the work on the Self of power requires a lot of effort, and such aesthetic construction of U.S. power has been made possible through resources provided by a wide variety of actors, including academic intellectuals, members of the U.S. congress, social organizations, bloggers, and other so-called journalists. The aesthetics are fueled by advances in technology that also spring forth their own categories of counterpower. Thus, in order to investigate U.S. power's aesthetic construction, such an inquiry requires us to visit a multitude of sites of aesthetic stimulation, including the processes of knowledge construction in the rise and fall of paradigms and American generations (chap. 1), discursive fields of contestation (chap. 2), the relationship between the academy and U.S. power (chap. 3), and the ambiguous mission statements U.S. soldiers find themselves pursuing (chap. 4). It is not limited to the present and future—such aesthetics in the form of the enamored relationship between Americans and U.S. power have been around at least since the middle of the Cold War, according to Andrew Bacevich (2005), and, depending on which classical realist thinker one consults (see Niebuhr 1932: 98–103), probably since at least the late 1890s and the Spanish-American War.

Further, while I do not think this book is specifically crafted in a way that will be digestible to what might neutrally be titled the "military establishment," I am quite confident that the (post)modern American military will recognize as obvious my arguments about the impact of aesthetic insecurity as it appears in the form of the case illustrations used especially in chapters 3, 4, and 5 of this book. Thus, the lessons of ambiguity in warfare and the temptation to right an aesthetic appearance of a wrong may serve as pragmatic warnings about the mission of the U.S. soldier. Most vividly, the argument that power operates in "the dark" through ambiguous operational objectives and that such operation inevitably entails costs, found especially in chapter 5, implores the military to redefine its relationship to the battlefield and its enemy.

Finally, this investigation also should give ethical pause to policymakers who use what can only be described as trite, even cute, but irrevocably *vague* slogans when justifying the use of lethal force in the name of the so-called national interest. Such slogans have permeated the U.S. cultural and political landscape over the past eight years: where a war, like a sporting event, could be "lost," a withdrawal from a battlefield was titled "defeat and retreat" or a policy of "cut and run" that showed a lack of "fortitude" or "will." The absurdity of it all reached a new plateau in 2008—when the chairman of the Joint Chiefs of Staff warned that a *timetable* or timelines for withdrawal of U.S. forces from Iraq would be "dangerous" but that the "strategic goals of having time *horizons*" should be sought.[8] Nevertheless, the move toward such seemingly semantic distinctions indicates, however, the *aesthetics* of such terms themselves as used by those who execute security policies, as well as the symbolic and strategic purchase such aesthetics can have in international relations. The analysis in this book suggests, in various ways, that abstract words help veil the operation of power with aesthetic content, but they also create, eventually, further vulnerabilities.[9]

Let me end this section of the introduction by stating the purposes this book *does not* serve. First, I wrote it not in the name of clarifying the ambiguities of power or as a polemic against power writ large but, instead, taking the ambiguities of aesthetic power as functional entities, to delineate what effects followed from them and what practices such ambiguity made possible. It is not a book, then, that seeks to propose an alternative "plan" to save the world from its own demise. As I mention in

8. This was stated in a 19 July 2008 interview of Admiral Michael Mullen on *Fox News Sunday* (available at http://www.foxnews.com/story/0,2933,386843,00.html).

9. I am reminded here of words used by Kerouac's *On the Road* character Carlo Marx, in his advice to his fellow beatniks about the dangers they were ignoring as they embarked on their cross-country trip: "The days of wrath are yet to come. The balloon won't sustain you much longer. And not only that, but it's an abstract balloon" (2007: 121).

this book's conclusion, I find such problem-solving and progressive moves to be an increasingly problematic style of writing that was all the rage for a previous generation of IR scholars. But it is not my style—for this book falls far short of providing an alternative order or even ethos that leads us to a more just and equitable account of human relations. I remain, instead, firmly resolved that the ethos of the Self that produces aesthetic creativity and transgression also contains built-in quandaries that emerge when the Self wanders into dark areas of operation. Our only resolution is to wipe the canvas of the Self clean, revealing it, so that an insecuritized U.S. power is forced out of the darkness in which it has so problematically operated over recent years, if not decades. For even if no pause by power results from the recognition of the aesthetics of insecurity—for it is not in the nature of those who celebrate movement and action to pause for anything—there will be costs. Aesthetic transgressions are a fact of power's operation, whether policymakers recognize it too late or not.

The Study of Aesthetic Insecurity, Power, and Counterpower

While the full contours of what I mean by the terms *security, insecurity, power,* and *counterpower* will be developed in chapters 1 and 2, I here briefly sketch what they mean, for the reader's convenience, in order to demonstrate how the field of IR already contains the theoretical tools to study this important topic. Security, as several scholars have identified, is an essentially contested concept, or ECC (Smith 1999: 106; Fierke 2007: 34–51). Fierke defines an ECC as a concept that "generates debates that cannot be resolved by reference to empirical evidence because the concept contains a clear ideological or moral element and defies precise, generally accepted definition" (2007: 34). The field of critical security studies has advanced the case that narrow definitions of security—for example, the defining of security in military terms versus other forms (environmental, health, or identity)—are a political position. The what of security is also contested alongside the who of security. Are we dealing with humans (Hampson 2002), society (Buzan 1993; Wæver 1993), the nation-state, a region (Buzan and Wæver 2003: ch. 3), or the broader global context?

Even the most preponderant conceptualization of security has been a site of definitional frustration. Wolfers (1952) claimed that national security as a "political formula" had been invoked quite consistently but was in danger of losing "any precise meaning at all" (481). This book in no way resolves the issue of conceptualizing security and insecurity, of course. Indeed, while I sympathize with Wolfers's frustration at the use of the term *national security* "without specification," as it "leaves room for more confusion than sound political counsel or scientific usage can afford" (1952: 483), I instead assert that *power is made possible and exists through the ambiguous space.* Such ambiguity is destroyed in the moment when the idealized "narrow cause of the national self" (ibid., 482) is revealed. The archaic definition of aesthetics—"the study of the nature of sensation"—is quite telling here.[10] As it more resembles an art than a science, security perhaps can be just as easily understood using aesthetics—the "feeling" one gets from a work of art, for example, is like "security," necessarily ambiguous and individualized but also a basis for action. When idealized beauty is compromised, insecurity follows. Since idealizations are by their very nature unattainable, this means that insecurity is quite common.

Yet Wolfers does provide two useful reference points. First, he notes that "security, in an objective sense, measures the absence of threats to acquired values, in a subjective sense, the absence of fear that such values will be attacked" (1952: 485). Remarkably, this conceptualization of security as the absence of threats to values remains a largely viable one in the field of critical security studies (see Fierke 2007: 14). We can modify this position slightly when we recognize how the aesthetics of the Self is a value on its own and consider which effects this makes possible. Second, Wolfers also identified a disproportion between power and security, where "some weak and exposed nations consider themselves more secure today than does the United States" (1952: 485 n. 4). In a like fashion, I investigate in this book why the United States would act so quickly—and, at times, comprehensively—against instances of aesthetic challenge. Instead of acquiring a sense of increased safety and physical security, the case illustrations demonstrate how increases in power lead to an inflated sense of Self manifested through an addiction to motion. Such senses and addictions cannot be sustained without periodic ruptures. This is what security as an "absence" implies.

What is meant by power? This book understands power in a multifaceted manner. As counseled by a recent expository on the subject (Barnett and Duval 2005), it sees power not only as the ability of A to get B to do what it otherwise would not (Dahl 1957: 202–3) but also in terms of

---

10. http://dictionary.reference.com/search?q=aesthetic. I am indebted to Jon Carlson for this insight.

a centralized body's internal capacity to perceive its ability to operate upon its own self-image, as well as influence others and determine outcomes. This process is attended to through both the acquisition of materials and their diffusion into an ontology of the Self. For instance, the voluminous mainstream literature in security studies assumes that for nation-states, material and strategic power helps to ensure survival (Waltz 1979: 126–27). Yet as power is acquired (presumably for this purpose), it becomes part of the routines of the nation-state's existence. Groups seek to acquire more power than is necessary, so that power becomes an end itself (Niebuhr 1932; Mearsheimer 2001; Schweller 1994). Power here is about more than ensuring physical survival—it can constitute the identity of the actors that have acquired it.[11] Such power may serve as a basis for influence exerted over Others, but it is also vitally important for the ontological security of the Self , the securing of self-identity through time.

I began this argument in my previous book, stating the following when discussing how physical capabilities influence an agent's sense of freedom and thus also paradoxically increase levels of agentic anxiety:

> More capably "powerful" states are somewhat imprisoned by their ability to influence more outcomes in international politics, and in this sense these capabilities, rather than allowing these states more freedom to act (as their acquisition is intended to accomplish) [instead] compromise this sense of "freedom." (2008a: 69)

At that time, using Giddensian sociology, my position was that all social agents, including states, sought out continuity for their actions—social routines that provided meaning and simplified chaotic life so that it became tolerable. Yet it is in seeking out this continuity that anxiety emerges (ibid., 48). In that book, I advanced the argument, albeit minimally, that this greater "freedom of choice" makes the more powerful incredibly anxious and therefore vulnerable to manipulation. One is reminded of Orwell's (1962: 95–96) sahib when he turns "tyrant," as recorded by James Scott.

> It is his own freedom that he destroys. He becomes a sort of hollow posing dummy . . . For it is a condition of his rule that he shall spend

---

11. Diplomatic and military historian Andrew Bacevich avers, in the opening page of his recent book *The New American Militarism*, that "the global military supremacy that the United States presently enjoys—and is bent on perpetuating—*has become central to our national identity*" (2005: 1, emphasis added). I discuss Bacevich's study in more detail in chapter 5.

> his life trying to impress the "natives" . . . He wears a mask and his face
> grows to fit it . . . A sahib [hegemon] has got to act like a sahib [hege-
> mon]; he has got to appear resolute, to know his own mind and do
> definite things. (Scott 1990: 11)

This conceptualization of power is productive (Barnett and Duval 2005) in that it exists and permeates through a field of operations. Such a conceptualization presumes constraints, of course, because power can become centralized into certain bodies, but it also presumes a freedom within this field: "If power were never anything but repressive, if it never did anything but say no, do you really think one would be brought to obey it?" (Foucault 1980: 119). Further, as I discuss in more detail in chapter 1, Foucault saw this field of relations as "mobile, reversible, and unstable" (1989: 441). While freedom helps bring about these relations, it also entails self-defeating effects as well. Or, as Andrew Bacevich (2007) suggests in his most recent work, freedom allows for power to find its limits; it produces possibilities but also crisis.

I am hardly the first IR scholar to recognize the inherent vulnerability of ontological power. Jutta Weldes has most persuasively argued, along similar lines, that because U.S. leaders—in general and, in her specific study, during the Cuban Missile Crisis—feared "appearing weak," U.S. identity was constructed "not only in opposition to the external other of secretive and aggressive totalitarians but in opposition as well to an internal feminine Other defined as weak, soft, complacent and self-indulgent." This produced "for the United States a pervasive and inescapable credibility problem" in which U.S. leadership "demanded continual demonstration" (Weldes 1999: 46). In an attempt to "project strength," these leadership narratives instead generated perpetual insecurity.

We can locate even more resources for excavating the intersections between aesthetics, security, and power within the broader traditions of feminist IR theory. Feminist IR has provided us a very expansive analytical terrain to examine power, positing, like Foucault, that power manifests in a field of relations and relationships. These relations occur in a multitude of spheres—from the family and household to society to the binary constructions created by scholars. Wherever these relations occur, they rationalize, or make normal, forms of public behavior (masculine) and relegate other behaviors (feminine) to particular private spheres. For decades now, feminist scholars have demonstrated an "interest in the uses and consequences of power," uses and consequences that include the way in which gender roles are constructed (Enloe 2007: 100). Feminist analysis provides a template for how we might go about uncovering the implicit forms that power might take, and it furthermore asks us to reflect on how power operates even through our own "knowledge claims" (Ackerly and True 2008: 694). There are four sets of additional insights that prove instructional for a critical investigation into aesthetics and power.

First and foremost, feminist theorists outside of the field of international relations have demonstrated the disciplinary implications of notions of beauty and aesthetics. Naomi Wolf's 1991 book, *The Beauty Myth*, is arguably the most well-known argument in this vein. Even though women had made inroads into professional, civil, and political spheres throughout the twentieth century, masculinized notions of how women should look continue to play a particularly nefarious form of social control. Although it may not be characterized as such, this particular work by Wolf views gendered forms of discipline in postmodern terms in many ways. For instance, take Wolf's assertion that certain "fictions" of gendered relations "transformed themselves" during this period and that because the "women's movement had successfully taken apart most other necessary fictions of femininity, all the work of social control once spread out over the whole network of these fictions had to be reassigned to the only strand left intact [the image of female beauty], which action consequently strengthened it a hundredfold" (1991: 16). What Wolf shows is that the myth of female beauty appropriates an ideal that perpetuates female insecurity. Yet in the foreword written for the 2002 reissue of the book, Wolf also discusses the possibility of a male beauty myth emerging, where "men of all ages, economic backgrounds, and sexual orientations are more worried . . . than they were just ten years ago" (2002: 8).

Moving to gender analysis in IR theory, we find several more important insights. The binaries exposed by feminist analysis (strong/weak, rational/emotional, public/private), constructed markers that have "serve[d] to empower the masculine over the feminine" (Tickner and Sjoberg 2006: 191), demonstrate how masculine identity exists in a relational form to femininity. What this means is that "powerful" masculine behavior, such as the emphasis on narratives of "strength and will" found in U.S. foreign policy crises (Weldes 1999), depends on an internal feminine other, which must be constructed as "weak" and submissive. Again the relational context of gender produces particular roles for organized bodies of power, including nation-states. It does not matter whether these roles are actually embodied, in total by "individual men and women," but rather that these are "ideal types" (Tickner and Sjoberg 2006: 186). As I discuss in this book, aesthetic identity itself embodies an ideal—a cosmetic notion of how a subject wishes to look. As these idealizations operate at the level of organized power, and as notions of masculinity can satisfy notions of aesthetic representation, aesthetic power can be said to be inherently related to and even vulnerable to the representations of it by counterpower. For instance, the notion that power's authority stems from its apparent orderly, rational, and confident posture (a posture associated with masculine notions of behavior) can instead be represented as anything but in a counterpower event. The highly emotionalized nature of aesthetic power—discussed in chapter 1—can also be thought of as the highly emotionalized substance of the masculine subject. Counterpower thus challenges the arbitrariness of the supposedly "natural" gendered role distinctions.

In a related vein, feminist analysis has also uncovered how the meanings of gendered roles are not constant through time but must be "produced and reproduced through symbolic and cultural practices." This is not so far removed from the "aesthetic representation" of power that I explicate in this book—and this will become especially clear in chapter 1, where I outline several of the "strata" of an aesthetic subjectivity of power, one that, like the meanings of gender and masculinity, draw on a variety of "cultural artefacts" (Hutchings 2008: 101). Such power gains its purchase only when it is in operation—its "physique" becomes most impressive when exercised. In this sense, feminist analysis has demonstrated how the state itself can be problematized as a masculinized social construction, one that is also dependent on the reproduction of particular cultural practices (Peterson 1992), which can serve to "justify" the hierarchical nature of nation-states vis-à-vis other social groups or associations. This position delineating how cultural representations create meanings for action is one I obviously share by writing a book on aesthetic power.

Finally, in a manner that intersects with the thrust of the analysis delineated in chapter 3, on academic-intellectual parrhesia, feminist analyses have proposed more careful attention to making power as "the researcher's subject, and not the researcher power's agent" (Ackerly and True 2008: 699). Feminist analysis, by excavating how particular concepts that permeate the field of IR (the state, foreign policy, identity, etc.) are gendered, due largely to the historical predominance of men in the field during the twentieth century, ask us to be particularly attune to how we as scholars are implicated in the field of power relations we study.

The one slight departure I take here with feminist analysis, both as it stands in social theory and as it has been utilized in IR, is that the masculinized itself is inherently insecure—that, furthermore, the expectations for particular performances required by masculine aesthetic subjectivity lead to, even themselves necessitate, actions that are physically, ontologically, and emotionally costly. Having to "live up" (or down) to this hegemonic masculinized ideal type (indexed by the "male beauty myth" that Wolf posits is now with us) creates fragility for the masculine subject.

More conventional studies, I think, intuit the possibility of perpetual expansion leading to increased levels of security, even if they delineate such insecurity produced by economic, geographic, or military conditions. Paul Kennedy's (1987) well-known theory of imperial overstretch asserted that the expenses of economic and military preponderance eventually outstrip the benefits great powers accrue during their "rise" to power. More recently, John Mueller (2005: 209–10) posited that imperial expansion generates its own insecurity and decline. Mueller writes about how Soviet communism "eventually collapsed of its own weight and lack of appeal and of the failure of its misguided, even romantic, worldview." Conventional accounts of imperial expansion recognize how such extension is "more the result of pressures in the periphery that lead to *unintended, unanticipated* political developments that generate reactions that pull great powers more deeply into the politics of other polities" (Nexon and Wright 2007: 267, emphasis added; see also Macdonald 2004: 9). Yet while these accounts largely draw on the material strain that maintaining a great power or imperial global position entails, they still recognize that the *ambitions* of great powers outrun their capabilities, creating these unintended consequences for decline. In other words, while these studies provide very detailed scope conditions through which (in their case) imperial or great power collapses, we might think of these unanticipated developments in another way—as power's exposed spaces. The more deeply extended (spatially, aesthetically, and/or ideologically) power becomes, the more vulnerable it is to such micropressure. As a result, there are opportunities for positioning counterpower within these conventional understandings.

Because counterpower is a micropressure, an unlimited event that can happen at any time and from any direction, it cannot be predicted and therefore preempted. Counterpower is "unbearably light, in that it is always possible," and "it is [therefore] unlimited" (Simons 1995: 82).[12] At any time, a challenge to aesthetic power can serve to stimulate it to engage and work upon its own sense of aesthetic integrity.

Why term this "counterpower" and not a rival form of power in and of itself? Surely actions or techniques that force a centralized body of power to *seemingly* do what it otherwise would not or act in a way it otherwise would not act constitutes power as well. While I articulate counterpower more elaborately in the following chapters, there are two immediate points in using the term *counterpower*. First, counterpower is not necessarily forcing a targeted power to do what it otherwise would not, in the sense that the coercion here originates instead through movement away from a previously delegitimized Self. Second, counterpower, understood in a microsense, is displayed as moments, styles, words, or images. These manifestations are infinitely small and light—they do not contain the material or strategic force necessary to maintain influence over a body, nation, group, or organization over time. Counterpower cannot itself distribute resources, taxonomize actors and actions, synthesize, coordinate, or routinize. Such displays derive their influence, alternatively, from the power that they engage. In this book, such power is American. But it is my assumption that counterpower can engage other forms of aesthetic power because, again, the process of aesthetic self-creation is nearly ubiquitous. In other words, while certain organizations are more inoculated than others from counterpower, all (individuals, groups, states, and transnational actors) contain an aesthetic subjectivity that can be disturbed by counterpower. The overturning of any of these groups, however, must come from another contending power (nation-state, world state, etc.), not from the counterpower challenge itself. That said, in conjunction with a series of power challenges, counterpower may serve to help unravel the Self of a power, which can be manifested in both productive and debilitative ways.

A final preliminary issue to consider is *where* this process occurs. Aesthetic insecurity in centralized bodies of national power occurs among and through several levels. On one level, it is, of course, the individual who is insecuritized, whose imagination is disrupted by the de-aestheticized image that counterpower helps bring about. On another, it is the *ontology* of power itself that is disrupted. Here, the idea that power can

---

12. It is not clear whether Simons intends for this to be a play off the Milan Kundera novel *The Unbearable Lightness of Being* (1984), although for those who have read this work, there are some parallels between Kundera's Nietzschean-inspired novel and the vulnerability of a power that searches for an aesthetic integrity that is never there.

get "others to do what they otherwise would not"—that it, in a sense, has *control,* or, more precisely, the ability to organize and then synchronize the Self's operations—is troubled. The disruption here occurs over how power sees itself, how it wishes to be seen by others, and the techniques it uses to re-create these scenes. This confidence in the ontology of power is, while not shattered, nonetheless disturbed in moments of counterpower. Thus, security-seeking behavior following a counterpower engagement is understood no longer (as it is in Wolfers) as the *retention* of values but, rather, as the attempt to recover an aesthetic integrity that was *perceived lost.* As the empirical illustrations reveal, such an aesthetic totality of power most likely never existed to begin with but was idealized. Precisely because it was idealized, even romanticized, rather than objectively realized, the centralized body of power was set up for this rupture long before it occurred. By seeking out aesthetic integrity, power facilitates its own vulnerability.

The Plan of the Book

The argument of this book is developed through two analytical chapters (chaps. 1 and 5) and several chapters (chaps. 2–4) that illustrate specific forms of counterpower. Chapter 1 builds the core concepts of the aesthetics of power and counterpower. It conceives of aesthetic "power" as composed of several strata—what I term the *psychological,* the *rhythmic,* and the *imaginative.* Both the classical realists of the twentieth century and recent post-structuralist work provide helpful building blocks to this argument. Aesthetic power and its vulnerability can also somewhat be understood with reference to the emphasis on action found in the vitalist philosophy of Carl Schmitt, and it finds further expression through the embrace of motion and strength found in the work of one of the founding fathers of the American neoconservative movement, Norman Podhoretz. Yet while this emphasis on the decision in vitalism has the stated intention of creating a more secure environment for the Self, it instead leads to pockets of vulnerability that arise because the aesthetic of power that had been operating "in the dark" is revealed in decisive action. My argument draws on several theoretical inspirations, including the aesthetic perspective provided by John Dewey, the recent engagement of aesthetics in the field of IR theory, and its main theoretical resource in the form of Foucault's aesthetics. Chapter 1 also introduces the "generational" analysis of the aesthetic images of power (especially U.S. power) and draws on various streams of such inquiry as located, for example, in the work of Kuhn on the cycles of scientific revolutions. This overview of generational analysis reveals how the production of knowledge itself is embedded with generational conflict and that shifts in this production are also impacted by aesthetics. Generational analysis also shows how technology not only (re)fuels generational imaginations of aesthetic power but also serves as one more avenue through which counterpower can dislodge power.

The next three chapters outline practices of counterpower as seen through a variety of case illustrations. Chapter 2 develops two levels of what I call *reflexive discourse,* the dialogue of one actor that insecuritizes a materially powerful target into acting according to the latter's sense of aesthetic integrity and self-identity. These two levels are influenced, in turn, by two accounts of social theory. Level 1 explicates how the ontological security of a powerful actor is challenged by reflexive discourse and that this can be understood using some insights from Giddensian sociology. Level 2 explicates how reflexive discourse includes elements of what Foucault titles "flattery"—tactics I discuss as "bundling" and "self-flagellation." Chapter 2 uses certain comments made by Jan Egeland during the context of the 2004 Asian Tsunami and the U.S. reaction to those comments to illustrate these two levels of reflexive discourse. Yet while flattery can have productive effects, in that it can help assuage the damaged ego of a powerful actor (as it did in the case of the United States following Egeland's initial "stingy" comment), flattery can also be what Foucault calls a "mendacious discourse" that *artificially* inflates the "Self" of the target. Chapter 2 proceeds through a brief illustration of flattery's "mendacity" in the form of the dialogue that members of the Iraqi National Congress used to convince the United States to act on its behalf leading up to the Iraq invasion of 2003.

Just as an aesthetic Self can be constructed via a "mendacious discourse," so, too, can it be broken down and stripped of its artifice through truth-telling. Such frankness is captured by the concept of parrhesia, the counterpower practice developed in chapter 3. The chapter begins by exploring Foucault's development of parrhesia and outlines the general contours of truth-telling in politics and the study of politics. It then analyzes some general conditions that constrain truth-telling in politics. The chapter then specifies two possible exercises of modern-day parrhesia—cynic parrhesia (as used, for instance, by transnational terrorist organizations) and academic-intellectual parrhesia. Cynic parrhesia is understood with reference to Osama bin Laden's 2004 address to the United States in the days immediately preceding that year's presidential election. Academic-intellectual parrhesia is examined through the case illustrations of the U.S. academy's participation in or against the construction of U.S. national security narratives during World War I, the Vietnam War, and the recent Iraq conflict. These illustrations explicate the "seductive" qualities of power that serve to inhibit intellectual parrhesia.

The final configuration of counterpower explored in this book is titled "self-interrogative imaging," which refers to the distribution of unfavorable images that represent (and re-present) the state being targeted. Chapter 4 explores the role of images in constructing, deconstructing, and de-aestheticizing the U.S. Self, as examined through the case illustrations of the My Lai massacre (1968), the U.S. withdrawal from Saigon (1975), and the more recent U.S. War on Terror. The latter cases analyze the images emanating from the contractors killed in Fallujah in April 2004 and the photographs depicting the torture of detainees that occurred at Abu Ghraib prison.

Chapter 5 takes stock of these illustrative insights, situates them with existing accounts of power, and briefly overviews how power has been configured in several categories of IR analysis. The chapter helps explicate the spatial and temporal frame of counterpower by casting it as a form of transgression, a recognition of the limit of the Self only when it has been crossed. Transgressional counterpower is contrasted with several "transcendental" and "transitional" accounts of power. The chapter addresses how a "transgressional" approach to power can contrast with but can also complement transcendental and transitional approaches to power.

The concluding chapter begins with an analytical defense for counterpower, with some propositions on how counterpower can be broadened in future studies. While the examples drawn throughout the book's case illustrations stem from U.S. national and international experience, the wider implication is that all bodies of centralized power—individual, substate, state, transnational, suprastate, transeconomic, and so forth—construct their own aesthetic bases for legitimacy and can therefore be challenged by counterpower. Counterpower admittedly rests on some ethically contentious positions. The "art of uncertainty" that is at the heart of counterpower analysis can be seen as a full-frontal assault on principles of order, continuity, and stability—principles that were part of the Hobbesian epistemological project (Williams 2004). Thus, the final sections of the book advance an ethos that may minimize the aesthetic vulnerability of power. This stems from Foucault's observation that the Self of the individual can be created as a harmonious ethical subject—

and thus that power, even narcissistic power, can still explore a "style" that seeks to minimize interference into the lives of others. One of these readjustments, the championing of restraint, is presented with reference to a diverse body of philosophers and theorists, from certain Stoics to twentieth-century classical realists. It argues for what I title *acupunctural formations,* a form of productive but localized human relations that can exist within the field of power relations. Acupunctural formations and the principle of restraint both reject the notion of comprehensive solutions. Instead, the ethos sketched in the conclusion seeks out pockets of freedom within a pervasive global field of power relations, pockets that may or may not be emulated more expansively.

ONE

Aesthetic Power
and Counterpower

This chapter advances an account of aesthetic power and counterpower. It begins by defining power as an aesthetic subjectivity of a centralized body of individuals, groups, nation-states, or transnational organizations. Again, this is not intended to be a full-frontal assault on the typologies already introduced—at least not in an analytical sense. Instead, the purpose of this chapter is to propose that aesthetic subjectivity obtains in power. It stems from several resources, spanning from John Dewey's work on the "experience" of art to the Foucauldian observation that subjects create the Self as a work of art but do so removed from or sometimes in opposition to comprehensive moral codes. The process of self-creation inevitably entails an engagement of the aesthetic as a communal experience; subjects look inward for guidance on the construction of the Self but also do so within a community of fellow citizens. Power is therefore consolidated not necessarily through the relations with an Other external to this community but, rather, intrinsically with a centralized Self.

This is not, however, an argument advanced solely on the terrain of ideas—acquired material resources are necessary. They are used not only to further survival, moral purpose, or the self-identity of collective bodies but also to continuously re-create the Self. Returning to the "makeup" metaphor discussed in the previous chapter, centralized bodies of power utilize resources to apply a cosmetic image to the operation of power. These resources are disseminated throughout the various agencies (or "applicators") to coordinate the movements of a centralized body—in a national context, this would include bureaucracies, branches of government, the military, science, and art. As a result, this aesthetic construction helps regiment the legitimacy of power.

It accomplishes this regimentation, however, by individualizing power—organizing it from the individual up to the centralized body. Aesthetic visions of power are engaged in a multitude of fashions by the individual—the openness to interpretation that defines the aesthetic allows the individual to internalize the Self of power. What makes power so attractive and legitimate for these subjects is the freedom to choose the meaning of this beauty. While theorists have established an ability of the nation-state to extend a "political power beyond [its spatial limitation]" (see Rose and Miller [1992], quoted in Löwenheim 2007a: 205)—shaping citizen choices for travel abroad with warnings, for example—the conceptualization of power here assumes that it is an attraction constituted within and between the confines of individual subjects.

Let me also provide here a bit of background to the role aesthetics plays in political and moral philosophy and justify why I choose John Dewey and Foucault as my philosophical building blocks of the aesthetic, as opposed to Kant, who is the inspiration for a substantial amount of aesthetic analysis. In *Critique of Judgment,* Kant sought to identify the process by which humans can judge something to be beautiful and agree with one another on that judgment. Though we can never know if others truly concur about a common object of beauty, the ability to make a judgment and to communicate that subjective judgment to another individual is evidence for Kant that we are rational beings (Kemal 1999: 287–88). What is important for Kant is not the knowing about the objects themselves but the "nature of our perception of them" (Bleiker 2001: 513). Therefore, it is not the object of art itself but, rather, the judgment as a form of assented communication that possesses a universal validity. The faculties of human judgment regarding other forms of cognition allow for us to claim, according to Kant, that our judgments about what is beautiful also are universally valid.

While this view by Kant is by no means ignored throughout the remainder of this book, the aesthetics of insecurity finds a bit more inspiration from the philosophies of Foucault and Dewey (among others), because of their emphases on the ability of art to work back upon the subject and their mutual rejection that any "ground" or scientific knowledge of the Self can be established. They also confront the notion that humans contain a universal will toward rational judgment, especially when confronted by the aesthetic. Instead, subjects continually transform so that judgment is always contingent in time and place. Or, as I discuss in chapter 5, a transgressional limit, when breached, shocks precisely because the subject is always unsure of its "true" Self. The Self's ambiguity and indeterminacy—the lack of an essence—are its most vulnerable and thus (counter)powerful qualities. A primary value the current chapter wishes to advance, then, is that such aesthetics of power contain a paradox. They of course allow power to operate because of the disciplinary nature of what I will term its three "strata," but by inflating the Self of power through such aesthetics, idealizing it into a more heavy body that cannot be sustained. Aesthetic construction, therefore, facilitates a vulnerability to the "weightlessness" of counterpower noted in the introduction.[1]

This position is elaborated further in section I of this chapter by examining aesthetic power's three strata—what I term the *psychological*, the *imaginative*, and the *rhythmic*. Section II situates this understanding of power by reviewing several vibrant accounts of the aesthetic in both broader social philosophy and more recent IR theory, with a special focus on Foucault's understanding of the Self's work on the Self. What is most instructive about the aesthetic is the sometimes traumatic process that follows a break, or rupture, as I also explore in section II. Section III pivots from these insights to articulate the function of and fixation on action found in vitalist ideologies, as examined through the work of Carl Schmitt and neoconservative commentator Norman Podhoretz. The aesthetic as it is realized through motion and action in these accounts has the stated intention of creating a more secure environment for a community or Self (a "show of force" or "demonstration of will"); it instead leads to pockets of vulnerability that arise because the aesthetic of power that had been operating in ambiguity becomes revealed through such action. These revelatory instances shock or disrupt the three strata of aesthetic power, and this is one fashion in which vitalist ideologies contain their own self-defeating logic.

Section IV concludes the chapter with a final assertion regarding the manner in which generations renew the Self of a political community. Through the creative capacities of individuals and, in certain instances, a democratic public, generations create and re-create the Self of community power. Generational analysis helps make intelligible the politicized conflicts that arise between intracommunity generations in a wide variety of realms, spanning from the production of knowledge (seen

---

1. Note especially the quote by Simons that counterpower is "unbearably light, in that it is always possible" (1995: 82), which was presented in the introduction.

through Kuhn) to the making of foreign policy. As generations rise and fall, they fuse with the power of a community to shape new visions of a community Self in often reactive (to a past generation) ways. The special role of technology in these cycles is discussed at the conclusion of the chapter.

I. Three Strata of Aesthetic Power

In a book that makes the case for the aesthetic subjectivity of power and how such subjectivity can be problematized by counterpower, it is necessary to provide a systematic conceptualization of power itself. What follows here are what I term the three strata of aesthetic power. I prefer the term *stratum,* rather than *component,* to refer to the *imaginative,* the *rhythmic,* and the *psychological,* because it may conjure some of the other contexts in which the term *stratum* has a meaning.[2] The individual functions of each strata are discussed in turn, before a brief exposition on the manner in which they work co-constitutively to build an aesthetic subjectivity of power.

Psychological

The import of a psychological stratum of power has long been anticipated by a variety of theorists writing about power and has been most persuasively advanced by Hans Morgenthau and other classical realists. Morgenthau's famous dictum declared political power to be "man's . . . control over the minds and actions of other men" ([1948] 2006: 30). Later, Morgenthau asserted that "power, like love, is a complex psychological relationship" ([1967] 1970: 243). Power is also "a quality of interpersonal relations that can be experienced, evaluated, [and] guessed at" (ibid., 245). Power engages the psychological when it is reduced to its most concealed form, as Foucault posited in *Discipline and Punish* when he wrote of the disciplinary society's propensity to "bring the effects of power to the most minute and distant elements" (1977b: 216).

Other classical realists have advanced this claim that power is much

---

2. One of these contexts is geological—strata are layers of rock that must be excavated in order to recognize their intrinsic qualities, but when the side of a rock formation is revealed, the layers appear in all of their levels. Another is anatomical and biological—skin, for instance, has layers. Jon Acuff (2008) discusses identity construction in similar terms but instead uses the more colorful and innovative metaphor of a "many-layered pastry" or, even more specifically, a mille-feuille.

more compelling when it is concealed—when the possibility of resistance is not even entertained.[3] I have referred to these classical realist emphases on "prudence-as-stoicism" and "prudence-as-restraint" in my other work (Steele 2007b),[4] but in regard to the specific stratum of psychology, it is important to note here Niebuhr's points on how social groups, "among whom a common mind and purpose is always more or less inchoate and transitory," maintain cohesion (1932: 55). The primary way that power can maintain its hold over the "masses" is, for Niebuhr, through a psychological connection established between the individual citizen and the idea of the nation (1932: chap. 5). Likewise, we can recall here Robert Cox's (and Gramsci's and Machiavelli's) metaphor of power as a centaur, half man and half beast, "a necessary combination of consent and coercion" (1983: 164).[5] Individual subjects are involved with power and submit to it because it fulfills their psychological needs, especially on the rare occasions it is exercised, and perhaps especially when it is violently expressed.

The use of power, furthermore, is emotionally satisfying. Such exercise activates, to paraphrase Reinhold Niebuhr, our "anarchic lusts vicariously" (1932: 93). The most extreme version of this psychological attraction to such exercise occurs, of course, as Erich Fromm posited, in a fascist regime, which "arouses and mobilizes diabolical forces in man which we had believed to be nonexistent, or at least to have died out long ago" (1964: 7). The society in which man inhabits contains a dual function—it suppresses, but it also creates (ibid., 13). But a psychological stratum in aesthetic power assumes an inverse effect on the "man on the street" the more coercion is exercised—for if physical power needs to be deployed more often, the likely cause is a *decline* in its psychological stratum of influence. In such cases, aesthetic power's subjectivity becomes an object—it travels from a concealed place to a revealed vision. Counterpower can assist this revelatory process, as I discuss later in this chapter.

One result of using a psychological stratum in aesthetic power is that, by definition, it problematizes rationality. It makes the contrary assump-

3. Reinhold Niebuhr posited that "the police power of the state is usually used prematurely; before an effort has been made to eliminate the causes of discontent, and that it therefore tends to perpetuate injustice and the consequent social disaffections" (1932: 35).

4. This is a position that I return to in the conclusion.

5. See also Morgenthau's famous statement, in his opening chapter of *Politics among Nations*, that "a man who was nothing but 'political man' would be a beast, for he would be completely lacking in moral restraints" ([1948] 2006: 15).

tion that social action is essentially complex and difficult to reduce to rational formulations (Morgenthau [1967] 1970: 242); therefore, the attempt to remove complexity from models of IR "obviates the possibility of theoretical understanding" (254). It is, however, not only impractical to remove such contingency and uncertainty through the rationalization of IR—if we remove such uncertainty (or, in a psychological sense, such anxiety or "dread") from our models, we are removing a primary *engine* of power: "the element of irrationality, insecurity, and chance lies in the necessity of choice among several possibilities" (254). It is ambiguity itself that facilitates the need for action (or motion), aesthetic construction, and emotion to obtain a pull on the groups (citizens and social movements) that help reinforce power's legitimacy.

Imaginative

Power's productive influence also depends on connected individuals (in a nation-state, these would be citizens) imagining that its operation is ethical and even beautiful. The imaginative stratum is intuited through certain post-structuralist accounts of power and action. We might begin with Lacan's (1977) work, as others have in more detail (see Debrix 1999: esp. 249 n. 5; Edkins 1999: esp. chap. 5). For Lacan, identity is developed, in part, during that stage when the child can see his or her image in a mirror, when the child can recognize himself or herself and notice how the movement of the body in the mirror can be controlled. The subject's identity is formed in relation to this image through a series of fantasies—fantasies that the infant is this total, this smooth, this collected image that appears to him or her in the mirror.[6] This is, thus, an *imaginary* sense of control, yet this imagination is necessary for the child to form a psychological continuity with his or her social environment. As a result, the child interprets that he or she may have the ability to control more (rhythmic synchronization), including the environment that surrounds him or her. That the child cannot in reality do so is not of concern here; rather, the focus is on the continuous tendency of the human subject to look for this smooth continuity in its environment and on how such a search is an illusory, tragedian drama of the human being. The imaginary order is a necessarily narcissistic one, and it accords to the sketch of aesthetic power that I provide in this book—self-centered, ego-

---

6. "The role of illusion in the formation of the ego is crucial. A continuing search for an imaginary completeness is one of its outcomes" (Edkins 1999: 91).

tistical, and, importantly, vulnerable to disruption by the Lacanian notion of the "real."

What protects the Lacanian subject from this disruption is the symbolic order, where we analyze the role that language plays in processing the image for the human. How can we recast the Lacanian orders as representations of the counterpower process? The subject's dependence on the imaginary is made vulnerable because this basis is erected through image fantasies, but it is dislocated through the real. The protective veneer of the symbolic order will eventually "interpellate" the image that challenges the Self, but it is in those moments when there is a suspension of the symbolic and when the image itself is too (counter)powerful to be characterized that power becomes disrupted. The image defies symbolic and even imaginary control. Recognizing it as a threat, the symbolic intervenes upon the real and classifies it, embedding it with disciplinary meaning. The events evoking correlated sets of images that are addressed in subsequent chapters represent those moments of the real—moments where power was briefly dislocated and symbolic signification was suspended; where power operates unattended in an ambiguous space; and, in certain cases, where re-actions by power against such images took place.

When discussing imaginative power, however, the narcissism of the "mirror stage" is deepened. For imaginary power, which interprets through the lens of Self the meaning of external events, each action that occurs "out there" *must* say something about the Self. The Russian attack on Georgia in 2008 or the current climate crisis, devoid of their own non-U.S. contexts, are interpreted by the U.S. subject as saying something about the U.S. Self. Yet when power goes looking for reflective images in others, it is vulnerable—it cannot predict which images of the Self will emerge, since it is looking everywhere for evidence of self-gratification and is bound to be let down (down from the idealized Self).

But why would power go looking for its reflection? According to Lacan, "the mirror stage is a drama whose internal thrust is precipitated from insufficiency to anticipation—and which manufactures for the subject, caught up in the lure of spatial identification, the succession of phatansies that extends from a fragmented body-image to a form of its totality" (1977: 4). The problem for the Lacanian subject, if we think (as I do) that it indexes power's search for aesthetic totality, is that these images are fragmented because the gaze of power looks upon *an* image only briefly. It may resurrect that image on occasion to accord to its fantasies, but the understanding of self-interrogative imaging advanced in chapter 4 sees such images as unexpected events. The narcissism of imaginary power becomes routinely dependent on these images and, in so doing, loses the ability for self-definition. In other words, power, which prides itself on control and the control of others, has an inherent *lack* of control over which images do or do not emerge in the mirror stage, which thus loosens its grip over aesthetic self-integrity. In the words of Edkins, "as a result of the mirror stage, the ego's mastery of its environment will *always* be illusory" (1999: 91).

We might also consult the works of recent post-structuralist IR theory, as evidenced by James der Derian (1990) and David Campbell (1998). The U.S. Self is perhaps an obvious choice for my conception of power's subjectivity as imaginative. Campbell states,

> If all states are "imagined communities," devoid of ontological being apart from the many practices that constitute their reality, then America is the imagined community par excellence. For there never has been a country called "America," nor a people known as "Americans" from who a national identity is drawn . . . Arguably more than any other state, the imprecise process of imagination is what constitutes American identity. (1998: 91)

For America, "absence," rather than "presence," leaves it without any moorings at all, *dependent,* in the words of Campbell, on "representational practices for its being," for its sense of Self. For Campbell, the need to maintain these practices makes America insecure. What Campbell titles the "globalization of contingency" and indeterminacy forces the United States to imagine itself, but always in his reading as a product of difference, in relation to an Other. By making the American state a "body," an object, American identity is actively constructed against dangers abroad and risks within. We might also add that the dependence of the United States on the *outcomes* of these representational practices makes it even more vulnerable, as they are hardly univalent. For der Derian, the processes or practices of surveillance and simulation show how "power is here and now, in the shadows and in the 'deep black.' It has not trouble seeing us, but we have had great difficulties seeing it" (1990: 304). Such power, he notes, derives its influence from its ability to represent truth (305).

One process that allows for power to maintain its disciplinary function is, of course, surveillance. Yet surveillance, one premise of which is to gather "intelligence" on adversaries for the ultimate purpose of security (national or otherwise, as there are various levels of surveillance), is dependent on a material and a social reality that makes power even more vulnerable than it might at first appear. To begin, in a national context, the agencies that collate information are themselves organizations with their own routines, or "standard operating procedures." Although such agencies are at the service of the state, they have their own goals, making power "factionalized" into organizational processes (Allison 1969). Despite attempts to gain centralized control over these agencies—evidenced by the recent U.S. Intelligence Reform and Terrorism Prevention Act of 2004, which appointed a director of national intelligence—built-in and fragmented organizational routines are extremely difficult to centralize.

Second, "paranoia" can result from surveillance because it produces inherently ambiguous images and data. This is the case with any form of surveillance. Where ignorance can sometimes be bliss, surveillance produces scenes that facilitate even more ambiguity and multivalent suspicions. Imagine a scene where a private investigator arrives at our door— he carries with him a picture of our significant other entering a restaurant. We cannot quite tell when the picture was taken or even the name of the restaurant he or she is entering, but perhaps most striking is that our significant other *appears* to be walking into the restaurant with a stranger. Even if a friend of ours looking at the same picture told us this was *not* the case through (à la Kant) a "rational" judgment of such a picture, we still may not believe this. The point here is that even in situations of friendship or love, we can resort to paranoia in the face of an image that makes multiple interpretations possible. What produces the paranoia, the insecurity, is our own imagination, concerning even or perhaps *precisely* someone we trust or something that we love (nation), if we have idealized that person's or thing's beauty. When faced with even the possibility of betrayal, we experience the beginnings of a rupture in how we think about that person or thing. Surveillance makes us more vulnerable than we would be otherwise.[7]

Surveillance—and the imaginative capacities it produces—can lead to problematic security strategies or tactics as well, as one example illus-

---

7. This does not imply that certain forms of surveillance can never be envisioned in positive or comforting ways, of course. Like any exercise of power, such surveillance contains the capacity to both disturb and soothe, to destroy and produce. See, for instance, the work Jack Amoureux and I conducted on a human rights panopticon, which saw nongovernmental organizations as serving a panoptic-function for liberal hegemonic power in their surveillance of genocide (Steele and Amoureux 2005).

trates. One of the practices by U.S. forces as they entered Iraq during the first months of Operation Iraqi Freedom was the decision to leave discovered weapon depots both unguarded and undestroyed. An insufficient number of troops necessitated leaving these sites unguarded, but why did they not detonate the weapons? One estimate put the amount at a million metric tons, all of which were conventional materiel and many of which would be looted and used by the growing Iraqi insurgency in the manufacturing of roadside explosive devices. As Ricks (2006: 145–46) discusses, prewar intelligence collectively assumed that Iraq possessed weapons of mass destruction, and the assumption was that some of these depots contained biological and chemical weapons. This was, of course, faulty information, and yet it produced a situation where the apparition of WMDs was depicted as reality. This positioning ultimately made U.S. power more vulnerable.[8] Morgenthau also recognized the problem with imagination and power. He asserted that while rational desires could be satisfied, the "will to power" could not. Such a will to power had a "transcendent goal" that was reached only in the imagination but "never in reality," as "the attempt at realizing it in actual experience *ends always with the destruction of the individual attempting it*" (Morgenthau 1946: 194, emphasis added).[9]

Where is the imaginative stratum of power exercised? Landmarks or iconic geography, for one, help constitute the aesthetic subjectivity of national power, as evidenced by nationalistic songs such as "America the Beautiful."[10] Again, Foucault's brief work on heterotopias is useful here—as these sites are places where, as Oliver Kessler notes, narratives and "geographic markers" collide (2008: 19–20). According to Foucault, heterotopias are sites of "counter-arrangements in which all the real arrangements . . . are at one and the same time represented, challenged and overturned." These are sites of function—they have "the power of juxtaposing in a single real place different spaces and locations that are incompatible with each other" (Foucault [1967] 1997: 352). Furthermore, they possess a duality of possibility: while "creating a space

---

8. A second form of imaginative stratum of power is illustrated in chapter 4, with reference to the recent U.S. embrace of torture as an interrogative technique.

9. Notice as well the overlap between this observation, which speaks of the problems with imagination, and the discussion in chapter 5, via Fromm and Liska, regarding the interrelationships between the transgressional and the transcendental.

10. Noteworthy in this context are the connection between the (at least rhetorical) support by the National Socialist Party in 1930s Germany and its ability to foster nationalism or the connection between ecology and what Bruggemeir (2005) and others title "crypto-nationalism."

of illusion that reveals how all of real space is more illusory, all the locations within which life is fragmented" (354), they also "have the function of forming another space, another real space" (1997: 356).[11] A heterotopia is not a "utopia" but a real space where the Self of power is experienced, refracted, and ultimately reinforced.

Let me be clear, as I am not positing the geographic markers or other historical landmarks to be "un-real." On the contrary, when citizens visit one of these sites, they are facing the "real," but with their own sense of imagination—they must interpret and reflect on the ideal in the face of the real. They must imbue the marker with meaning. Again, because power contains an imaginative stratum, two citizens viewing the same marker can individualize the meaning of power's beauty, and each of these meanings may prove difficult to intersubjectivize.[12] Beauty can defy articulation, and even if it can be articulated, individuals may feel reticent about doing so, if they wish to keep this subjectivity a secret (an assumption I make, again, contra Kant).

The experience of seeing the "majesty of purple mountains" or a national monument, however, is becoming more difficult because of recent economic trends—trends that, combined with technological developments, may accelerate the efficacy of aesthetic "simulation." As der Derian also discusses, simulation is represented through "spectacles" that go further than reality, into the realm of the "hyper-real" (1990: 299). As a result, the real can no longer be distinguished from the ideal (Baudrillard 1995). Consider how the touring by Americans of these sites will be reduced in coming decades due to the petroleum-dependent U.S. economy, rife with increased fuel costs and a nearly dysfunctional national public transit (i.e., speed rail) system. Hence has come the appeal of so-called staycations, where families spend their summers nearer to home. The aesthetic power of geographic or historical landmarks and ostensibly the "idea" of the nation then must be simulated—moving from heterotopias to pixeled utopias—clipped, Photoshopped, JPEGed, and downloaded, rather than contextualized by the docent or travel guide. Thus, the possibility therefore exists that the imaginative stratum

11. On heterotopias, see also Salter 2007.
12. Anthony Lang briefly discusses, via Arendt's *Human Condition*, the three realms of active life: labor, work, and action (Lang 2008: 52). While Lang focuses on action to further develop his notion of agency (see also Lang 2002), a formulation I discuss later in this chapter, his definition of the artifacts of work (2008: 53)—"that activity which results in good that outlive us; creation of buildings, art and crafts that are not consumed but remain after individual human lives pass away"—could also be considered here sites or markers for imaginative power.

of aesthetic power will become more prominent, reinforced, and individualized in coming years.

Rhythmic

Finally, aesthetic power contains a rhythmic stratum, a stratum that becomes clearest in Dewey's aesthetics, discussed later in this chapter. To aver that aesthetic power contains a rhythmic stratum involves four meanings. In one sense, the rhythm of power is illustrated by the aforementioned routines of organizations—"standard operating procedures" of bureaucratic groups that execute a policy based on a particular scenario (Allison 1969). What element of aesthetics exists here? In the (idealized) *synchronization* of these specialized agencies, in their coordinated movements once a "crisis" erupts, they are like an offense in football—each member has a specific task (routing, blocking, passing, etc.), the totality of which culminates in a perfectly executed "play" (Allison 1969: 702). In a second sense, as presented in the last section of the current chapter, cycles of generations also provide rhythms to the various "paintings" of the aesthetic value of the state.

In a third, more structurationist sense, the "rhythm" of power is seen in and through the routines agents perform that attend to their senses of Self. Put another way, the Self of a body of power is constituted through routines that entail structures—structures that work back upon the agent by providing meaning to the latter's existence. Such structures filter out the chaos of everyday life by reducing it to the most predictable elements. The argument, drawing on Anthony Giddens's structuration theory (1984, 1991) that these routines obtain at the level of groups, organizations, or nation-states, has been advanced by myself (Steele 2008a) and several others (Huysmans 1998a; Kinnvall 2004, 2007; Mitzen 2006; McSweeney 1999). Certain treatments of ontological security (see Steele 2008a; Kinnvall 2004) focus on the role that narrative plays for agents in making sense of an individual or group Self. As Lang (2002) has noted, narratives help to articulate a self that is in constant action. In terms of rhythm, a narrative smoothes out and provides coherence to the active self. Further, this element of coordinated rhythm is also evident in the emerging work on time and international relations (Hanson 1997; Hutchings 2007; Hom forthcoming), which has demonstrated how sovereign power (among other forms) helped discipline groups and individuals through the advent of "clock time." Time itself is a social construction that explicates the routine, the rhythm, and the expectation and anticipation inherent in models of secular, economic, and political development.

Finally, the rhythmic stratum is apparent through ceremonial rituals or other communal celebrations of power—such as national holidays like the Fourth of July, where many firework displays throughout the United States are not only conducted with patriotic music in the background but (like that of organizational processes) synchronized with this music on local radio stations. But the spontaneity of such a celebration also reveals its intensity—viewers cannot predict what they will "feel" when they see a firework, but they know it will be spectacular nonetheless.[13] Air shows are similar sites, where spectators can witness both the technological power of a nation's or organization's aircraft as well as the *skill* of pilots.[14] The rhythmic stratum is on full display here, as the pilots adeptly synchronize their movements, titled "aerobatics," to entertain the spectator.

The Collective Strata of Aesthetic Power

We can configure many ways in which these three strata of aesthetic power work in unison, but let me provide a general schema of scenarios for their collective synthesis. The general scenario scales up an aesthetic construction of power's subjectivity with the rhythmic, seen through routinized movements of the body, and a narrator who imbues these movements with an ambiguous but compelling meaning for action, important for identity. The rhythmic stratum co-constitutes the psychological, as these routines provide us not only an anchor of ontological security of the Self but also emotionally satisfying narratives of what the Self is and what it means in time and space.[15] During traumatic (emotional and psychological) ruptures of the Self, a reparative narrative can also reinstall the idealized Self (Edkins 2003; Steele 2008a: 56–57) by speaking around the opening or rupture. As Dewey's analysis discussed later in

---

13. The temporal element here is important to power, as Adorno points out: "Many artworks of the highest caliber effectively seek to lose themselves in time so as not to become its prey . . . Ernst Schoen once praised the unsurpassable noblesse of fireworks as the only art that aspires not to duration but only to glow for an instant and fade away" ([1970] 1997: 28).

14. A search for air shows in 2008 in the United States alone on www.airshowbuzz.com produces 216 different air shows that spectators can attend. But the phenomenon of the air show extends throughout the world, as that same Web site's search engine attests.

15. Besides my own work on ontological security, the relationship between psychology and identity is investigated in Brian Greenhill's (2008) recent study.

this chapter suggests, rhythms as public rituals intersect with the imaginative stratum as they provide a template for individualizing publicly expressive ceremonies into idealized visions of a group Self.

The psychological stratum connects to both in important ways. Imaginative ruptures of the idealized Self must be processed through the psychological stratum as emotionally intense reactions, and those reactions impact the rhythm of daily existence. A terrorist attack, for instance, creates a hole in the imaginative stratum by challenging the smooth, imagined notion of power (control over one's own image). It also manifests extreme psychological and social-psychological emotions (fear, sadness, melancholy, and, of course, anger), and it most assuredly impacts the rhythms or routines of daily existence, such as travel. In this way, a terrorist attack impacts aesthetic power much more abruptly than the physical consequences it engenders.

II. Aesthetics and Power

In this section, I focus on Dewey's perspective on aesthetics as well as some of the emerging aesthetic IR accounts, with special concluding focus on Foucault's aesthetics, an account that I suggest most accurately anticipates the Self of power. The affinities between Dewey and Foucault were first probed by Richard Rorty, who commented on how the two "agree, right down the line, about the need to abandon traditional notions of rationality, objectivity, method and truth" (1982: 112). This places both, again, against Kant's aesthetics. Jim Garrison has more recently noted the similarities between the two on the process of self-creation. While both Foucault and Dewey begin with this process, "Dewey looked outward [toward the Self in the political community] while Foucault looked inward" (Garrison 1998: 112). Indeed, the purchase of Dewey's insights is that we can apply the notion of aesthetic insecurity to the democratic community of the U.S. Self, a community that engages in "*creative* participation." This will become most useful when we understand the potentials for the reform of the democratic Self during a de-aestheticized moment of counterpower, as evidenced in the following three chapters.

Dewey: Experience and Emotions

John Dewey, in *Art as Experience,* sought to disaggregate the idea that art is an object and, instead, asserted art's proper position within a cultural setting as something a society experienced. Deweyean scholar Robert Westbrook avers, "At the heart of Dewey's aesthetics was an effort to break down the barriers between art and the rest of experience and to trace the continuities between the work of art and the doings and undergoings of everyday life" (1991: 390). This requires a temporally and spatially *contextualized* approach to art as experience. The Parthenon, for example, "has esthetic standing only as the work becomes an experience for a human being," namely, in how it came to be an "experience" for the Athenian citizens and thus in what those "whose lives it [Parthenon] *had entered* had in common, as creators and as those who were satisfied with it" (Dewey 1934: 4, emphasis added).

It is because art has been separated from "reason" and science, from everyday experience, that we fail to recognize just how important aesthetic experiences are to our everyday lives. In fact, as I discuss via Kuhn in section IV, the aesthetic exists even in the supposedly objective field of scientific inquiry. Dewey's assertion is that spirit and matter are combined and that science and art both depend on an aesthetic experience. The binaries in which we separate form from function, style from substance, and spirit from matter are shattered in Dewey's aesthetics. Quoting poems in their entirety, Dewey asks, "Did anybody who felt the poem esthetically make—at the same time—a conscious distinction of sense and thought, of matter and form? If so, they did not read or hear esthetically, for the esthetic value of the stanzas [rhythms] lies in the integration of the two" (1934: 132). *This* is the aesthetic experience for Dewey—such experience provides unity between matter and form, between style and substance (37).

There are two further imports to note here in Deweyean aesthetics. One, conveyed briefly in previous sections of this chapter, is that aesthetic experiences are both public and individual, and both of these loci relate to Dewey's deployment of "rhythms." In the community, the aesthetic experience is fostered not only by the role of public monuments but through public rituals and celebrations, which are "expressive objects" that imbue collective meaning (Dewey 1934: chap. 5). By the bonds of community, "every intense experience of friendship and affections completes [themselves] artistically. The sense of communion generated by a work of art may take on a definitely religious quality" (Dewey 1934: 270). This intensity serves to glue humans to one another in this community and creates incentives for engaging the aesthetic integrity of it through productive forms of social critique (Garrison 1998: 126). The aesthetic experience brings the individuals into contact with their environment and even further melds them with their surroundings. The Self of the environment works on the individual self. Art is nothing until it is consumed through humans, or through "the living creature": "In an experience, things and events belonging to the world, physical and social, are *transformed through the human context they enter,* while the live creature is *changed and developed through its intercourse with things previously external to it*" (Dewey 1934: 246, emphasis added). Thus, the work of art that we see must be experienced through us, and in the process of working back upon us, it transforms our Selves, making judgments running narratives that are, from the moment they are uttered, highly contingent.

This binding quality—binding individuals to their community and to the work of art at a physiological level—is made possible through rhythm, which Dewey defines as an "ordered variation of changes" (1934: 154). Rhythm exists in nature, and humans acquire and then (attempt to) emulate their sense of rhythm from that source (148). Dewey elaborates,

> Experiences of war, of hunt, of sowing and reaping, of the death and resurrection of vegetation, of stars circling over watchful shepherds, of constant return of the inconstant moon, were undergone to be reproduced in pantomime and generated the sense of life as drama . . . The formative arts that shaped things of use were *wedded to the rhythms of voice and the self-contained movements of the body.* (148, emphasis added)

By participating in these rhythms, which are, once we emulate them, no longer "just" natural but, rather, "a matter of perception," the self has therefore "contributed . . . in the active process of perceiving" (163).

A minor drawback (for the purposes of this book) to Dewey's account is its lack of engagement with the implicit power of the aesthetic. He gives us glimpses that he recognizes such a possibility—"art is the extension of the power of rites and ceremonies to unite men, through a shared celebration, to all incidents and scenes of life" (Dewey 1934: 197)—but the purpose behind his argument lies elsewhere. Instead, one might notice that while "rhythms" of nature may have been reproduced by humans to provide life meaning—they have also served to discipline rather docile bodies (Foucault 1977a). What meaning is there in the rhythmic activities of the worker—such as the traffic jam, the lunch hour, boarding a plane, "tax season," and such? When such mundane experiences become (as Dewey casts them) aesthetic celebrations, we can still posit that power is "at work," underneath the facade of communal beauty.

Yet there is much to harvest from Dewey's aesthetics, and it is his insights into the role of emotions in the aesthetic experience that prove most beneficial to an account of counterpower. When beauty is challenged by being de-aestheticized, the emotional "rupture" is so intense in the audience that such a challenge inhibits clear thinking. Considering the recent proliferation of emotional analysis in IR,[16] it is apt to briefly audit the role of emotions in Dewey's aesthetics. Emotions are, in his aesthetics, not "simple and compact" outbursts but, rather, when significant, "a complex experience that moves and changes" the subject that experiences them (Dewey 1934: 41–42). An attachment to "the work of art" of the nation-state, for instance, presupposes a baseline of *affect.* Power becomes vulnerable when its aesthetic vision becomes overpristine. Art and images are able to disrupt and heal at the same time. When aggregated as something that excites, art "stirs up a store of attitudes and meanings derived from prior experience," and "as they are aroused into activity they become conscious thoughts and emotions, *emotionalized images*" (65). The emotion swamps the Self: "There is an explosiveness due to absence of assertion of control. In extreme cases of emotion, it works to disorder instead of ordering material" (70). Thus, the aesthetic experience is filled with emotion, and during ruptures, it works to "disorder" the agent who is experiencing art, including the art of power. That a community of individuals can work to try to repair this disorder is also assumed by Dewey, as well as several contemporary IR scholars who I now turn to. As readers will notice especially in the context of the case illustrations in the chapters that immediately follow the current one, this is an assumption that implicates the creativity and vulnerability of a democratic populace.

Recent Aesthetic Turns in IR Social Theory

Roland Bleiker's (2001) article on the "aesthetic turn," mentioned in the introduction, differentiated what he termed "mimetic" approaches to international relations, which sought to "represent politics as realisti-

---

16. See Crawford 2000; Mercer 2005; Hymans 2006; Ross 2006; Saurette 2006; McDermott 2004; Löwenheim and Heimann 2008; Sasley 2010; Sucharov 2006. Also see my discussion of this in my previous work (2008a: chap. 1).

cally and authentically as possible," from aesthetic ones, which "assume that there is always a gap between a form of representation and what is represented therewith," a gap that is (I would obviously agree) "the very location of politics" (509). Bleiker notes how even mimetic representations are invested with power, and in an insight that telegraphs the themes of Kuhnian analysis engaged in the last section of this chapter, he notes how "this power is at its peak if a form of representation is able to disguise its subjective origins and values" (515). Gerald Holden's (2006) review of aesthetic IR (also engaged in the introduction) outlined some of the many functions that IR scholars have used aesthetics for: (1) uncovering "solutions to global political problems," (2) investigating how "alternative narratives can disrupt conventional understandings of . . . territoriality or sovereignty," and (3) further developing feminist and postcolonial insights (801–2).

What aesthetic IR has been most useful for when we are investigating aesthetics and insecurity is revealing how a "shocking image" or "rupture" invades the routines of power. As Francois Debrix points out, for Slavoj Žižek (1989), the rupture is so severe that while a shocking image releases into a sublime image, not only does it preclude reinscription, "but it exists and matters as a haunting mark, a scarring piece of reality that denotes something that refuses to hide the gaping wound left by the initially terrifying sight" (2007: 138). Much of this work has examined the role of the sublime in international politics. A special issue of *Millennium* (2006) included several contributions to the topic. Antoine Bousquet, for instance, posits how "apocalyptic events" such as Hiroshima and the 9/11 attacks "ruptur[ed] our sense of continuity of time, thus forming a temporal break and omnipresent point of reference around which we subsequently *reinscribe* our historical and political narratives leading to the event and flowing from it" (2006: 741, emphasis added). While such reinscriptions are likely, these "time zero" events "offer the opportunity to . . . resituate our thought within history and outside the narratives that have so dominated Western consciousness" (ibid.).

Bleiker and Leet's (2006) work on the sublime distinguishes it from aesthetics and beauty, as something that "is linked to excitement and astonishment" rather than the "calm sense of harmony" of beauty (717). Yet the sublime is indeed "a break in the normal course of affairs," although, because of the distance with which we encounter a sublime occurrence, there is also "seemingly misplaced feelings of pleasure and delight" (721). The shocking image that Žižek argues leaves a "haunting mark" is, in Bleiker and Leet's argument, something that can instead be engaged in a healthy manner. Using insights from Nietzsche, the sublime image can instead be "displaced by delight when we become aware of our own role in *constructing* the scene around us" (729–30). This is a hopeful possibility, for as in Dewey's aesthetics, it implies how spectators or citizens become agents—they become aware of how they might become actors instead of audience members, and they help shape what they view.

In Debrix's book, the aesthetic regeneration of power occurs in problematically violent manners, stemming at times from what he titles "a peculiar form of sovereignty derived from Arendt's concept of agonal action" (2007: 114–15). Such action itself *is beauty*—even violence, properly expressing "passion, emotion, and heroism," becomes beautiful and "has no value or essence on its own, other than that which is represented and aestheticized every time the heroic agonal agents . . . perform their agonal and violent deeds" (ibid.).[17] This is a more daunting possibility, as the image of violence is retained and reinforced. Rather than stimulating repulsion, individuals become routinized and satiated only through further scenes of violence.[18]

Foucault's Aesthetics of the Self

I now turn to Foucault, whose "aesthetics of the self" is constructed from the "ethical relations to the self," which posits its fashioning without full recourse to societal guides or communal codes, emphasizing, in the process, creativity and innovation (see Osborne 1999: 46–48). It thus allows us a window into how power can be refashioned when stimulated through a counterpower event that focuses on the Self, focusing on power's responsibility to its own integrity.

Foucault posits that "the practices of the self thus take the form of an art of the self, *relatively independent* of any moral legislation" (1989: 458, emphasis added). An agent acting "moral" should thus not be equated with an imperative to uphold these societal codes or prescriptions. In another work (2008b), I have used this aspect of Foucault's mode of subjection to distinguish between internal and external modes of honor,

---

17. This is almost contrary to Adorno's view of aesthetics, where "the impression of ugliness stems from the principle of violence and destruction" (1997: 46).

18. Ritualized violence, then, would resemble, if not necessarily replicate, the destruction produced by the spectacle of fascist propaganda such as the 1934 Leni Reifenstahl film *Triumph of the Will*, which included scenes (synchronized marches, music, speeches, etc.) revealing all three strata of aesthetic power.

concluding there that the former, internal form exerts a more forceful pull on the subject than the latter.

This is a riskier position to take up in IR theory than it might first appear, considering the tension it engenders against assumptions guiding much English School and constructivist work, namely, that principles of the international community (ES) or intersubjective understandings or societal norms (constructivism) help shape nation-state behavior, identities, and the social or "moral purposes" that guide their actions.[19] My argument is that self-identity is a form of discipline and can be much more coercive than the disciplining function of social-communal standards, what some Foucauldian scholars have termed *command ethics* (Bennett 1996: 667).

While this conjures, in international politics, seemingly dire scenarios in terms of the inabilities of explicit international codes (law, conventions) or implicit norms or intersubjective understandings to constrain hegemonic power, the intrinsic mode of subjection can still produce a reactive process of transgressive reform. In fact, it is the precise contrast that this mode of subjection obtains against a command or external code that developed as the process Foucault outlines regarding moral action,

> in which the individual delimits that part of himself that will form the object of his moral practice, defines his position relative to the precept he will follow, and decides on a certain mode of being that will serve as his moral goal. And this requires him to act upon himself, to monitor, test, improve and transform himself. (1982: 28)

For example, at the time of this writing, it is politically advantageous in the United States for opinion leaders to eschew and ridicule international codes, laws, or institutions. Some of this probably stems from the "exceptionalist" identity of the U.S. Self. But besides expediency, another reason the U.S. power resists constraining itself with reference to international codes of conduct is because it sacrifices the U.S. identity status of "strength." Becoming dependent on such codes (even if they were originally instituted in part by the United States) now appears feckless. In other words, the reason why, to paraphrase Foucault, "the idea of

---

19. The term *social purpose* is found in Ruggie 1982. *Moral purpose* is derived from Reus-Smit 1999. Martha Finnemore, while using constitutive logic regarding agents and structures, still posits that the "purpose of intervention" is "bound up in other normative changes, particularly sovereignty norms and human rights norms," both of which are societal-level variables (2003: 57).

morality as obedience to a code of rules is now disappearing, has already disappeared [for Americans]" (1989: 451) is because the United States *resists* this code as a form of counterauthority to its hegemony.

Thus, the move to aesthetic identity, past constructivist and English School articulations of "moral identity," references these Foucauldian insights on the Self. Precisely because the Self's only recourse is to the Self, because the "self is not given to us," "we have to create ourselves as a work of art" (Foucault 1984: 351). The contours of this Self can never be ascertained. Our own being eludes us, although we perceive its "essence" to be ascertainable, hence the move toward self-creation (Garrison 1998: 116–17). Likewise, without recourse to epistemological norms, the construction of the Self resembles more of an art than a science (Osborne 1999: 46–47). This is what Foucault titled a "relationship to oneself" (*rapport a soi*), where the improvement of the Self comes through exercises (askesis) "such as self-interpretation, consciousness raising, dialogue . . . etc." (Tully 1999: 96). Foucault emphasizes the importance of learning these techniques that will aid in the art of living and in the care of the Self. These techniques are constantly sharpened through time, as the "care of the self" is "an activity" that takes place for all of life (Foucault 2005: 87–88). A Foucauldian emphasis on internal sanctions assumes not that there is *no morality* but that recourse to moral codes inhibits the ability of agents to create, innovate, and fashion their lives aesthetically. Instead, morality is "a matter of knowing how to govern one's own life in order to give it the most beautiful form possible (in the eyes of others, of oneself, and of the future generations for whom one could serve as an example)" (Foucault 1989: 458).

Foucault is indeed similar to Dewey in the sense that style and substance fuse into *one* aesthetic experience, but Foucault goes further by positing the aesthetic experience as a "provocatively self-involved" one "in which beholding a stylized [self-]reflection not only arrests the self in its gaze but leaves it *forever changed*" (Lamb 2005: 46, emphasis added). Furthermore, the artist is "his own spectator"; in fact, the urgency behind the artist's need to create art is that the artist wishes to learn more about himself or herself. Foucault states, "Why should a painter work if he is not transformed by his own painting?" (1989: 379). Moreover, Foucault tracks with Bleiker and Leek's perspective (previously noted) in that the artist recognizes *his or her own role* in constructing the work of art. Via Dewey, such self-knowledge launches avenues for groups of citizens to help in this fashioning of a collective Self. Thus, if we assess how this indexes to a materially powerful democracy like the United States (as I do), then the citizen has a role in creating the image, and the image at the same time transforms the citizen. Or, put another way, political agents craft these images for the purposes of power, and citizens of democracies can help in this fashioning that makes possible that power: such "glory cannot be dissociated from aesthetic value," for "political power, glory, immortality, and beauty are all linked at a certain moment" (Foucault 1984: 354). This fashioning can also, for Foucault, be a useful end in itself—we can step back and take pleasure at the sensuousness of the body's motion as a site of power. In fact, this is when we become most vividly aware of the body—for the individual, this includes such practices as "gymnastics, exercises, muscle-building, nudism and the glorification of the body beautiful" (Foucault 1980: 56).[20]

That said, none of this implies that the art of the Self occurs in a hermetically sealed jar, as the art is also connected, for Foucault, to the field of "power relations," mentioned in this book's introduction. It is within that field of "reversible" relations that the ethics of the Self's work on the Self occurs.[21] Some of these exercises for self-improvement are thus learned from society. Likewise, one purpose of self-creation is to create the context for further encounters, both with the other and with the Self, in order to "interrogate the circumstances [internal and external] that make that role possible" (Lamb 2005: 43).

Counterpower's Aesthetic Engagement

If one accepts that ideological power is not only material but also aesthetic in its construction, then such aesthetic visions can be challenged. The knowledge that this construction happens within the Self, without full recourse to international principles, does not mean that a singular power is oblivious to outside interpretations or presentations of it—on the contrary, it makes it ripe to be manipulated when its own aesthetic self-vision is challenged. Yet counterpower's challenges to aesthetic power centers are much more compelling when counterpower uses that inflated, narcissistic Self as its dominant analytical referent, rather than societal codes of "responsibility" or moral beauty. This is perhaps the

---

20. This, I believe, tracks with Jane Bennett's observation that such mastery exemplifies a "disciplined form of sensuousness" (1996: 654).

21. Barnett and Duval's taxonomy places such a field of relations in the form of "productive power," which they define as "the constitution of all social subjects with various social powers through systems of knowledge and discursive practices of broad and general social scope" (2005: 56).

most striking difference an aesthetic subjectivity of power maintains vis-à-vis the communal accounts of identity advanced by certain constructivists and English School scholars. It is when counterpower targets the Self of centralized power on *the latter's own self-constructed terms* that manipulation becomes not only possible but most forceful.

I understand counterpower in this book as *a micropressure that challenges the psychological, imaginative, and rhythmic subjectivity of power.* Counterpower is "micro" in a temporal sense; it has no real spatial or material quality. It is a style, a word, a set of words, a position, or an image. Counterpower works *within* the set of power relations it challenges. It prides itself on its spontaneity and innovative capacities, and its quickness is its strength.[22] By assuming that the nation-state is constructed aesthetically, as a work of art, I thus understand counterpower not as some "code," law, or counterplan—recognized by society and imbued with horizontal and vertical authority—but, rather, as a micropressure that may stimulate power to engage its own "art of living," since that art of living, as the Self's work on itself, is a "way in which people are invited or incited to recognize their moral obligations" (Foucault 1984: 353).

As suggested in the introduction, power becomes vulnerable, in my account, because of several configurations. First, when power's sense of aesthetic Self becomes inflated and taken for granted, it is exposed to de-aestheticization. Second, such manipulation of power can work in any direction—"good" or "bad"—in that the exercise of counterpower sharpens the targets' ability for "self-operation." This is what Foucault titles the "technology of the Self," where the aesthetic vision of the Self can be modified and reconstituted,

> techniques which permit individuals to perform, by their own means, a certain number of operations on their own bodies, on their own souls, on their own thoughts, on their own conduct, and this in such a way that they transform themselves, modify themselves, and reach a certain state of perfection, of happiness, of purity, of supernatural power, and so on. (2007: 154)

---

22. While I am hesitant to explicate revolutions or social movements as counterpower examples, it should be noted that spontaneous challenges to power are used as the exception, rather than the rule, of most security analyses. For example, Ole Wæver, in his famous 1995 chapter on securitization and desecuritization, notes that "only in rare situations—as during the 'Velvet Revolution' in Czechoslovakia, do we see moments—*almost seconds*—of a kind of self-evident representation of 'society' by some nonelected but generally accepted institution . . . It is *much more common* for a societal 'voice' to be controversial and only partly accepted" (70).

This leads to a third configuration of vulnerability—that the insatiable nature, or will, toward seeing the Self in action exposes a subject's power, what I discuss in the following subsection as a problem of vitalist ideology.

While Foucault's discussion of aesthetics seems like an obvious inspiration for the Self, it may appear to some readers a curious choice to utilize Foucault, of all analysts, to understand counterpower. Indeed, Foucauldian themes of the pervasiveness of power have generated the notion that Foucault took "contemporary society to be a scene of domination in which our forms of subjectivity are shaped without remainder by powers we have no effective capacity to resist" (Owen 2006: 130).[23] Yet this view of Foucault's early work is not entirely accurate. Take, for instance, what he stated at the end of his seminal work *Discipline and Punish* (largely understood as an all-consuming account of power) about how discipline "must *also* master all the forces that are formed from the very constitution of an organized multiplicity; it must neutralize the effect of counterpower that spring from them and which form a resistance to the power that wishes to dominate it: agitations, revolts, spontaneous organizations, coalitions—anything that may establish horizontal conjunctions" (1977b: 219, emphasis added). Additionally, Foucault asserted that his earlier work had insisted "too much on the techniques of domination" (2007: 154) and that he understood the word *power* as "short-hand for the expression . . . 'relations of power,'" relations that can be "mobile, reversible, and unstable" (1989: 441).[24] More than that, in such relations, "there is necessarily the possibility of resistance because if there were no possibility of resistance . . . there would be no power relations at all" (ibid.). Power can therefore dominate these relations, but it is never "in control." Its mastery, what made it "strong," "becomes used to attack it," and "power, after investing itself in the body, finds itself exposed to a counter-attack in that same body" (Foucault 1980: 56).[25]

Counterpower's influence depends on its spontaneity and its inability

23. Bleiker's 2000 book beneficially brings forward Foucault's view on power relations and how this makes possible a more "optimistic interpretation" (137).

24. See Simons 1995, esp. chap. 7. Simons states that Foucault, in this later turn, "remains confident that resistance is possible because power relations do not solidify into states of complete domination" (83).

25. Avid readers of Theodor Adorno might recognize how his immanent critique similarly assumed that "we have no power over the philosophy of Being if we reject it generally, from outside, instead of taking it on in its own structure—turning its own force against it" (1973: 97).

to be foreseen. For a brief illustration of this temporal element of counterpower, we might turn to the example of Cindy Sheehan that Debrix discusses in his *Millennium* article on sublime events (2006: 787–91). Sheehan's son Casey was killed in April 2004 while serving with the U.S. Army in Iraq. In the summer of 2005, the grieving mother went to President George W. Bush's ranch in Crawford, Texas, where the latter was vacationing. As Debrix details, Sheehan set up her protest in a ditch outside of the ranch, with the plea to speak to Bush to ask him why her son was killed in Iraq—what purpose did his death serve? What made Sheehan's visit to the ranch counterpower was that it was "unannounced" and "unwelcome" and that, after Bush continued to deny her a conversation, it led to a "spontaneous camp of tents" of fellow grieving mothers and war protestors around Sheehan's original tent (Debrix 2006: 787).

Indeed, this was, as Debrix titles it, an "interruption" that was an "abrupt mark" on how the war was depicted. It also disturbed the rhythm of the U.S. narrative that attempted to categorize the Iraq War as a "war on terror" (see Delehanty and Steele 2009). Casey Sheehan's death is perhaps more poignant because he was killed during one of the first Shia uprisings in the Sadr City area of Baghdad. It was a death at the hands of a domestic Iraqi population that had hardly been an ally of the Saddam Hussein regime, let alone an ally of Osama bin Laden's al-Qaeda. Debrix, however, rightly points out that events like the Sheehan protest "have no guarantee of political or democratic success," that they "may be captured, co-opted, or forcefully erased," while others can "actualize a critique of ideology and politics" (2006: 788).

But counterpower ends when it is captured by power, or, in Foucauldian terms, when it becomes "classified" or quarantined. The trajectory of Cindy Sheehan's image in the four years since Debrix wrote his article attests to this. After being widely hailed for this four-month protest, Sheehan began being depicted as an extremist, and actions by Sheehan certainly reinforced this perception, including her characterizations of Bush as a "terrorist" and her visit with and support for Venezuelan president Hugo Chavez, a protagonist of the Bush administration. Less than a year after Sheehan vaulted onto the world stage with her spontaneous act of counterpower, she was being interviewed by members in the media as follows:

> You lost a son in Iraq. We honor his service and sacrifice. But you've been traveling the world—Scotland, Spain, Venezuela, Ireland, Australia, Austria—how does that help the cause when, again, you're

around the world trashing the president, calling him a terrorist, calling him worse than Osama bin Laden. How do you honestly expect to affect change with those types of remarks? . . . Why go stand side-by-side by Hugo Chavez in Venezuela? Why do that? I mean, it sounds like—would you rather live under Hugo Chavez than George Bush?[26]

Because it derives its force from the element of surprise—because the insecurity of power is derived from the unpredictable—when counterpower enters the gaze of power, the latter dissects it as an element or object of knowledge. The further and further the message of counterpower travels from the time of the original surprise, the closer it gets to what Foucault would call the structure of "natural history," where spontaneous elements become observed and classified and provide "a certain *designation* and a controlled *derivation*" (1970: 138). The depiction of Cindy Sheehan, and even her selective reinforcement of that depiction, made it "clear" to many Americans that she was not simply a grieving mother or even a legitimate activist but a fanatic, one who would rather cavort with "anti-American" leaders such as Chavez. In the end, Bush never met with Sheehan, nor did he deviate from his consistently stale defense of the meaning of the Iraq War as connected to the events of 9/11.[27] The Sheehan example shows how difficult power is to permanently dislodge or transform. Again and instead, as I articulate in chapter 5, counterpower is a more limited and brief transgression.

Nevertheless, I do not wish to imply that powerful agents only act in selfish ways at the expense of external Others and that counterpower will only result in dangerous reactions by power. Counterpower's efficiency depends on power's aesthetic insecurity, but counterpower itself can possibly insolventize power to the point where the latter's reactions result in normatively fruitful outcomes. One need not assume, in the mold of a liberal idealist, that discursive, linguistic, or visual forms of counterpower will always produce progressive, transformational changes that

---

26. Norah O'Donnell, *Hardball with Chris Matthews*, 5 July 2006 (available at http://www.msnbc.msn.com/id/13735484/).

27. When asked, for example, about Sheehan's protest in a 25 August 2005 press gaggle, then deputy White House press secretary Trent Duffy did not even mention Iraq by name and instead stated, "The President has spoken continuously about the way he approaches this war, following September 11th, 2001. On September 14th, 2001, he stood at the National Cathedral and told all of America that this was going to be a very long and difficult war, and that there were going to be some very trying moments; but that because of what happened on 9/11, that we had to view the world in a different way" (http://www.whitehouse.gov/news/releases/2005/08/20050825-2.html).

will result in the betterment of humankind. But one must also not assume that power's inflated narcissism cannot be cajoled into productive possibilities as well. Indeed, "acting beautifully" and being "stylish" might demand that power take these productive turns on occasion, as I believe the Asian Tsunami illustration, presented in chapter 2, confirms.

Technology and Counterpower

As with many conventional accounts of power—the concerns expressed by classical realists during an "atomic age" being the most vibrant example (Herz 1959; Aron 1966; Morgenthau 1961; see also Russell 1991)—counterpower is especially sensitive to technological advances.[28] Morgenthau noted this in *Politics among Nations,* as modern war had become "push-button war, anonymously fought by people who have never seen their enemy alive or dead and who will never know whom they have killed" ([1948] 2006: 251). In addition, these advances in technology come packaged with the need for "technical experts," an elite group who must be consulted by those in power because of the inherently "esoteric nature" of technological power (Morgenthau [1964] 1970: 221). This development not only implicates power but leads to its own democratic deficit.[29]

However, while many conventional and innovative accounts have focused on the pervasive insecurity generated by advances in military technology, the current depiction of counterpower as challenging the aesthetic integrity of power sees informational flows and image dissemination as primary—rather than the potential for material physical destruction. In fact, in my view, it is the internalization of this conventional assumption—the dependence on material power—that has caused many security analysts and even state policymakers to be off guard during periods of aesthetic irruption. The counterpower of the image supersedes, in some respects, the short-term material damage of an event, as the illustration of the Fallujah case in chapters 4 and 5 suggests. It is only over time that such material destruction debilitates the physical security of power.

While technology can be used to further consolidate power, it can also dislodge and disturb it in several ways. Morgenthau mentions that

---

28. See also the work of Daniel Deudney (2007) on how nuclear weapons created the "one-worldism" movement.

29. I discuss this process of "expertization" in greater detail in chapter 3.

the co-optation of science by government for the singular purpose of technological advances that aid in national power actually decreases scientific creativity, "retard[ing] and distort[ing] scientific progress" ([1964] 1970: 228). Returning to der Derian, who sees technology as taking us from "reflection to reification" because we see "less of ourselves in the other," we might pose the alternative possibility that with technology, we see more and more of ourselves. Such a "mirror on the wall" forces us not to reflect but to re-create. If technology sustains power and if we operate under the illusion or belief that there "can be a single power or sovereign truth that can dispel or control the insecurities, indeterminacies and ambiguities that make up international relations" (der Derian 1990: 298), we are setting ourselves up for a very hard fall. For the increased quantity of countertechnologies contain the ability to expose power to be, if not out of control, at least aesthetically frayed at the edges and indeed insecure, ambiguous, and indeterminate.

Further, forms of technology (such as the Internet) accelerate the dissemination of information so that such information remains slightly "ahead" of a power attempting to classify and regiment it. Winograd and Hais relay the experience of Ruport Murdoch, a neoconservative media mogul who owns Fox News, someone who "may be, in fact, the last in a long line of media barons who have attempted to use their control of the distribution of information to direct the course of American politics and policy." "But," they add, "he has found that advances in technology [namely, the rise of the Internet] had *eroded the ability of those in power to stay there*" (2008: 152, emphasis added).[30] The Internet itself is a much more unpredictable medium than television or radio forms of communication and thus naturally evokes counterpower possibilities and realities. This is most plainly evidenced by the case illustrations in chapters 3 and 4, where the instantaneity of the Internet played a role in both the Cynic parrhesia of Osama bin Laden and the dissemination of images that interrogated the U.S. Self during the recent Iraq War.

A final point to make is that while the account of power sketched in section I of this chapter focused on its aesthetic elements, there is also evidence that the struggle *between* aesthetic power and counterpower occurs in aesthetic domains as well. Certain theorists see disruptive and

---

30. I return to this generational issue regarding technology in section IV of this chapter. At the end of chapter 4, I engage the possibility that we may have reached a point where technology has made power perpetually suspended and frail.

even emancipatory possibilities for art, where resistance could submerge into underground forms of artistic expression. This is, I believe, precisely what Roland Bleiker asserts as occurring in his account of "transversal dissent." Bleiker excavates several microsites of resistance—poetry readings, public toilet graffiti, music, art, and so on—which can bring forth, in limited terms, "new social movements" (2000: 205–6). Yet we might heed the warnings evident in Adorno's *Aesthetic Theory,* which takes as its basis the assumption of "absolute freedom in art" but nevertheless comes into "contradiction with the perennial unfreedom of the whole" ([1970] 1997: 1). Conventional art becomes the object of ridicule if it challenges power, if it becomes "pitilessly ideological"; a "solemn tone would condemn artworks to ridiculousness." So, to become unridiculous, conventional art resorts to the banal. Conventional art becomes aligned with—in fact, indistinguishable from—other "consumer goods." Adorno posits (among many other assertions) that power permeates the "culture industry" of a society, premised on what he terms the "pseudo-individualization" that causes society to collectively enjoy artistic outlets. What the individual does not recognize, of course, is that power permeates even the domain of art—it becomes a site of discipline, a marker where deviance is separated from convention (Adorno [1970] 1997: 39).[31] Thus the domain of possible spontaneity and resistance closes up.[32]

However, even without such microsites of resistance, what Adorno may overlook here is that the more consolidated and conventionalized the domain of art becomes, the more it idealizes (implicitly, even subconsciously) power or facilitates a vulnerability by lulling us into cliché-ridden boredom from genericized art such as elevator music. This is an idealization or normalization that is reinforced through the types of simulations and other imaginative practices I mentioned in section I. In the end, such idealizations or routinizations of power's subjective beauty cannot sustain their own weight. Counterpower, again a light, unpre-

---

31. Readers familiar with the currently popular U.S. reality show *American Idol* might find apt here Jeff Gordinier's description of it, which seems to echo Adorno's dire characterization of the culture industry: "*American Idol* was so perfect—an example of totalitarian kitsch at its most sophisticated. *American Idol* weeded out the freaks and churned out consistent, if uninspired, product . . . The message was: *Get with the program.* The objective was obedience" (2008: 90).

32. See also der Derian, who references Siegfried Kracauer's 1926 chronicle on the "emergence of a 'cult of distraction'" where the individual can go to seek "asylum from Weimar disorder, ornate spaces where the alienated Berliner could seek reunification through a new, totally imaginary, cinematic (yet organic) *Zeitgeist*" (1990: 299).

dictable force, uses those idealizations, rather than its own, to force out a rupture in the smooth aesthetic being of power.

III. Vitalism, Action, and Vulnerability

My purpose in this section of this chapter is to advance the argument that vitalist perspectives of politics, which are, in many ways, the celebrations of aesthetic power, contain self-generating logics of vulnerability. A dictionary definition of *vitalism*—"a doctrine that the processes of life are not explicable by the laws of physics and chemistry alone and that life is in some part self-determining"[33]—demonstrates the central importance vitalist philosophy places on action. Such love of action develops into an emphasis on aesthetic style via movement. Vitalist ideology sews its own seeds of collapse in its inherent dependence on the search for the Self's need to move against *something*, to act and react heroically. Further, I point out that the reason such emphasis on action obtains in these accounts is not necessarily because action successfully securitizes the subject but, rather, due to the vitalists' admiration for movement, specifically the intense celebration of the moment when the physique of power is revealed. Several vitalist accounts could be analyzed in this section,[34] but I focus mainly on an essay written four decades ago by a modern neoconservative author, Norman Podhoretz, as well as the work by and on the late German jurist Carl Schmitt.

Podhoretz's Vitalism

The emphasis on movement as an aesthetic is at the core of modern neoconservative philosophy. A telling essay written by Norman Podhoretz in 1963 illuminates the origins of this emphasis, back when neoconservatives still considered themselves American liberals—before the iconic moments of the 1960s fragmented Podhoretz's children's generation. Podhoretz and this essay in particular are engaged here for several rea-

---

33. See http://www.merriam-webster.com/dictionary/vitalism.

34. For a review of vitalism's development in the late nineteenth-century and early twentieth-century philosophies of various German conservatives, see Richard Wolin's (1992) work on Schmitt. As the notable work of George Mosse and others has demonstrated, aesthetics were integral to the operation of fascist regimes. See, in this vein, special issue, *Journal of Contemporary History* 31, no. 2 (1996). The current section advances the argument that the addiction to presentation as *movement* in vitalist ideologies, which include not only fascism but also other "regenerative" philosophies, makes the subject intensely vulnerable.

sons. First, he was and still is a very influential neoconservative author, serving as the editor of *Commentary* magazine from 1960 to his retirement in 1995. He thus represents one of the icons of an intellectual movement that has variously colored U.S. foreign policy since at least the Vietnam War. Second, Podhoretz's essay, written in 1963, predates much of his political activity and notable "break" with his left-wing allies in 1967.[35] This essay is, in many ways, a more intimate gaze into the key formative experiences of Podhoretz's childhood. Third, because it provides such a formative experience relayed as social commentary during a time when many white American liberals, like Podhoretz, were grappling with what they saw as the inherent tensions existing between liberal sensibilities, on the one hand, and the decadent "realities" of America, on the other, it provides an important but by no means sole example of the generational contexts that made possible the neoconservative movement.[36]

Two points of caution should be duly noted. First, I am not here trying to equate neoconservatism with Schmittean realism. While there has been some speculation that modern neoconservatives have been influenced by Schmitt,[37] several vibrant defenses of Schmitt have rightly pointed out that for all of the similar emphases on action and aesthetics between the two, there still remain important distinctions—most especially Schmitt's emphasis on politics as a "pluriverse" in contrast to the universalist logic of neoconservatism (Behnke and Bishai 2007: esp. 108–13). Second, I am fully cognizant of the uncomfortable subject of a white American speaking about his "Negro problem—and ours," as his essay is titled, but that particular topic and argument is not why I chose this essay. Rather, how Podhoretz demonstrates his argument is much more relevant to the intersection of aesthetics and action.

Podhoretz's essay relays the many confrontations he had with "Negroes"[38] while growing up as a Jewish child in Brooklyn, New York, in the

---

35. For a biographical note on Podhoretz, see *Contemporary Literary Criticism*, ed. Tom Burns and Jeffrey W. Hunter, vol. 189 (Gale Cengage, 2004), available at http://www.enotes.com/contemporary-literary-criticism/podhoretz-norman/introduction.

36. The oft-heard quote from another founder of the movement, Irving Kristol, was that a neoconservative was a "liberal mugged by reality" (*Two Cheers for Capitalism* [1978], quoted in Sidney Blumenthal, "The Ultimate Neoconservative Weapon," *Washington Post*, 9 October 1985, 3).

37. A good engagement of these can be found in Antoine de Benoist 2007. See also Behnke and Bishai 2007: 122–23.

38. Hereafter I omit the quotation marks, but let me assure the reader that I am only using the term Negro because that is the term Podhoretz uses in his 1963 essay, not because I condone or even agree with using this term for African Americans.

1930s. In a move that foreshadows the neoconservative disdain for intellectuals, Podhoretz's experiences were at odds with the theory that "all
Jews were rich . . . and all Negroes were persecuted." In his case, Podhoretz was continually "beaten up, robbed, and in general hated, terror-
ized and humiliated" by Negroes in his neighborhood (Podhoretz 1963:
93). On one occasion, Podhoretz and some of his friends were forced off
a ball field by a group of Negro kids, but not before a fight broke out;
and after being beaten, Podhoretz felt his "first nauseating experience of
cowardice" (94). On another occasion, Podhoretz was beaten by a "surly
Negro boy named Quentin" and the latter's brother after Podhoretz an-
swered a question in school that Quentin, when also asked, could not
(94). In yet another vignette, Podhoretz is surrounded and beaten by
five Negroes after a track meet where Podhoretz's (white) team was vic-
torious over several Negro teams.

On one level, what we might call the "Hobbesian," Podhoretz's neighborhood indexes international anarchy—a natural function in a realm
devoid of a supreme authority, a neighborhood where ethnic groups
must look out for their "own" and where they are, as he mentions in two
different places in the essay, "at war" with one another, where "only the
uniform" counts rather than the person (96, 97). But on another level,
what might be termed the heroic, this is more than just a typical fight for
survival—it is a fight of good versus evil. Podhoretz is punished for *good
behavior*, beaten by Negroes for getting an answer right in school and for
winning a track meet against a Negro team who was disqualified for
cheating. In the cases he relates, he is outnumbered (either 2 to 1 or 5
to 1), and outweaponed (Quentin's brother uses a baseball bat to beat
Podhoretz). In many ways, it represents a typical liberal quandary over
conflict, of which Morgenthau helpfully captured the essence when he
posited,

> Liberal condemnation of war is absolute only in the ethical and philo-
> sophical sphere and with respect to the ultimate political goal. In im-
> mediate political application, this condemnation is qualified and
> holds true *only for wars which are opposed or irrelevant to liberal aims* . . .
> When, on the other hand, the use of arms is intended to bring the
> blessings of liberalism to peoples not yet enjoying them or to protect
> them against despotic aggression, the *just end may justify means other-
> wise condemned.* (1946: 51, emphasis added)

Thus, conflict is not only justified but necessary to "protect" an enlightened enclave—such as a nation or, in this case, an individual like Podhoretz—from "despotic aggression."[39]

A further point Podhoretz attempted to make in this essay was that white America, especially liberals, were fleeing (i.e., in suburban flight) from confronting or interacting with the Negroes whose cause they had supposedly championed, and it was with this uncomfortable reality that white America needed to come to grips. Yet the purchase I see in this essay is not the sometimes incendiary dynamics of race relations or even its signification of Podhoretz's experiences to international anarchy but the particular Negro characteristics that Podhoretz "envies" (using that exact word).

For one, Podhoretz envies Negroes because of the freedom engendered by their dominance—they can enjoy *being* "free, independent, reckless, brave, masculine [and] erotic" (1963: 97). For another, "they were *tough*" (98). Such strength engendered a space for Negro freedom. But Podhoretz's "envy" extends into adulthood, for a quarter century after these experiences in Brooklyn, Podhoretz admires Negroes

> today [1963] for what seems to me their superior physical grace and beauty. I have come to value physical grace very highly, and I am now capable of *aching with all my being* when I watch a Negro couple on the dance floor, or a Negro playing baseball or basketball. They are on the kind of terms *with their own bodies* that I should like to be on mine, and for that precious quality they seem blessed to me. (99, emphasis added)

Now, as a scholar reading this in the late 2000s, I of course recoil a bit, for it is crude and culturally distasteful. But I also find it a fascinating glimpse into *how* Podhoretz synthesized action and aesthetics—"physical grace"—and further fused them with his praise for the "superior masculinity" (ibid.) of the Negroes from his boyhood memories.

There are, of course, many more modern examples of neoconservatism's vitalist emphasis on action[40]—the perspective is sometimes col-

---

39. While neoconservatism supposedly, according to Mearsheimer (2005), has both an "idealist strand" and a "power strand"—where "Wilsonianism provides the idealism, an emphasis on military power provides the teeth"—I think Morgenthau's insight, issued some sixty years prior, indicates that the so-called liberal ambivalence (see Rengger 2002) to conflict is highly conditioned and relative, rather than absolute.

40. See also Feith 2008; Bolton 2008.

loquially termed *movement conservatism*—but I think Podhoretz's essay bears special attention not only because of the time when it was written but also due to its blending of masculinity with eroticism, action (even violence) with beauty, and how they are idealized as *the blueprint* through which security in anarchy can be obtained. The Negro here is a *model of dominance,* but what perpetuates the dominance is not only strength but will, not solely material power but coordination and grace. This is largely the *representation* of and model for action and aesthetics that the modern neoconservative fetishizes and indexes to U.S. hegemony.

Schmittean Vitalism

The focus here on early twentieth-century German legal philosopher Carl Schmitt arises not only from the rapid emergence over the last decade of IR engagements with Schmitt's work but also because there are certain interesting elements of his vitalism that intersect with Foucault's aesthetics of self-creation, the latter of which is integral for understanding the vulnerability of the subjectivity of power discussed throughout this book. Huysmans has defined Schmitt's vitalism as "political life as an act of pure will" (1998b: 585). Like his conservative German contemporaries who were also somewhat inspired by Nietzsche, Schmitt offers an intensely critical view of the inherent nihilism resulting from Western values (Wolin 1992: 427). For Schmitt, politics had moved in the nineteenth century into a domain of "rationalization" and depoliticization, a point that foreshadows the conduct of generations (discussed in section IV of this chapter) in that generational change is predicated upon "shifting centers" (Schmitt 2005: 82). What is needed is an approach to politics and society that opposes such rationalization and routinization, with "life understood as a pure creative act of will not mediated by reason" (Huysmans 1998b: 582).[41] The need for the decision arises from the fact that in "spiritual spheres," there is an "ambiguity of every concept and word" (Schmitt [1929] 2006: 85). It is the exceptional condition that produces the decision—he who decides also decides the exceptional. It is in this defining moment, however, that the de-

---

41. The work by Erich Fromm also helpfully engages this creativity of the Self through ritualized action. Fromm argues that power maintains little purchase "in possession as such . . . What we use is not ours simply because we use it. Ours is only that to which we are genuinely related by our *creative activity* . . . only those qualities that result from our *spontaneous activity give strength to the Self and thereby form the basis of its integrity*" (1964: 261–62, emphasis added).

cision produces a rupture in the old order, which simultaneously produces a new order. Mika Ojakangas titles this the "double function" of Schmitt's central concepts (2006: 210).

Before discussing how these vitalist views become vulnerable to forms of counterpower, we might briefly consult the intersections between Foucault's views of the Self and Schmitt's vitalism, an intersection most impressively studied by Sergei Prozorov (2007), who advanced several common components, two of which I will touch on here. First, the emphasis on movement in both is apparent in Schmitt's emphasis on the decision replacing an old order and on Foucault's insistence on the productive capacities of intrinsic transgression (discussed briefly in the introduction). The subject emerges not prior to the act but in the act itself (Prozorov 2007: 231). The second intersection concerns Foucault's mode of subjection, where individuals are "the mode of subjection" or, to return to the discussion of the transgression noted in the introduction, the mode in which the subject engages the "space" left when the lightning flash dissolves (Foucault 1984: 353; Prozorov 2007: 231). This intersects with Schmitt's view that decisions must be made independent of transcendental moral codes (whether they are divine or human in constructive origin). As Prozorov astutely notes, a vitalist ideology like Schmitt's defies a grounded reference.

This vitalist view of the Self centralized by a "void which conditions [the subject's] being" (Prozorov 2007: 236) nevertheless exposes the Self of aesthetic power. In many ways, Erich Fromm anticipated this argument, as early as 1941, with the notion that the "lust" for power is "not rooted in strength, but in weakness." And what a weakness it is—exposed aesthetic power, which creates vulnerability, demands another exposure to close up the previous one. As Fromm further notes, the use of such creative power "is the expression of the inability of the individual self to stand alone and live. It is the desperate attempt to gain secondary strength where genuine strength is lacking" (1964: 162). We can add to Fromm's essentialist insight by noting how the conceptualization of power presented earlier in this chapter presumes ambiguity. Aesthetic rhythms derive their influence in their polyvalent abilities to work upon subjects in diverse ways—forcing the subjects to "connect the dots" themselves on the meanings of power. In other words, if power has an aesthetic layer, its compulsive logic stems not from a decision but from a "balancing act" between determinate substance (produced by the decision), on the one hand, and wholly indeterminate nothingness, on the other. The balance is ambiguity. Intersubjectivity here is organic, a bottom-up process that is generated and regenerated through time. The defining *decision*, however, removes this ambiguity—it objectifies what was once imagined. Because that which was imagined was idealized as beauty, the decision to essentialize that ambiguity is a decision to discipline imagination. While gaining influence through securitizing a threat, the decision instead removes the engine of power's aesthetic basis—the fantasy of the subject.

A further tension reveals even more precariousness in vitalism. What Schmitt and some of his contemporaries recognized as an exception becomes an exceptional condition that is itself normalized (Huysmans 1998b: 580); indeed, the rules themselves are dependent on the decision that ratifies and confirms them for society. Without it, such rules are translucent, even opaque—they are, in the word of Williams (2003: 517), "indeterminate." But what is routinized in the exception? To begin, the sudden "rupture" that justifies action is "manufactured . . . by fabricating an existential threat which provokes experiences of the real possibility of violent death" (ibid.). The heroic act done in the blink of an eye becomes a performance, judged not on its connection to confronting a security threat, if one exists, but on its artistic expression. This implies that a security threat (enemy) is not really there, is fabricated out of whole cloth. Yet it is not solely the threat that is manufactured; the Self (friend) must be established as well. Following a decision, the Self is also not really there, since the one identity of the Self that stems from the decision becomes isolated from the multiple identities a society could obtain. In noting this point, Williams accurately describes this as "reality . . . being denied" (2003: 520).

Schmitt and Podhoretz's accounts can be situated more broadly alongside social theories that emphasize action, especially in their philosophical (but not ideological) overlap with Arendtian notions of power. For Schmitt and Podhoretz, certainty, leadership, and strength all follow logically toward the action itself, an action that again demonstrates strength and authority and mobilizes a decadent public for further action. Similarly, in Arendt, as Anthony Lang avers, "action does not just create spaces and institutions for politics, it creates the agents themselves." Agents "realize" their sense of Self through action, or as Lang states, "once they [agents] appear on a public stage . . . human agents become a definitive 'who' as opposed to a 'what'" (2002: 12). This is simultaneously a collective, social act of the group. "Power," Arendt once wrote, "is never the property of an individual; it belongs to a group and remains in existence only so long as the group keeps together" (Arendt

1969: 43; cf. Klusmeyer 2005: 139).[42] All these forms of vitalism assume that agency exists in action, and the social need to act makes powerful collectivities much like coiled springs, ready to re-act.[43] So when the who becomes uncertain, when ontological insecurity is generated, the collectivized agent finds an incredible incentive to re-act in order to reorient their sense of Self, to reestablish the legitimacy of such agency.

This exposes another particular vulnerability in Schmitt's work in respect to the self-defeating logic of vitalist action. Far more vibrantly than simply sitting back and waiting for the exceptional moment, power goes in search of it—constructing it as the basis for "living": "in the exception the power of real life breaks through the crust of a mechanism that has become torpid by repetition" (Schmitt 2006: 15). We are, so to speak, "dead" until these moments and ruptures happen. The capacity to be spontaneous realizes the full Self of our being (Fromm 1964: 260). Moreover, for Schmitt, if all we have to fear is death, we have no life.

> A life which has only death as its antithesis is no longer life but powerlessness and helplessness. Whoever knows no other enemy than death and recognizes in his enemy nothing more than an empty mechanism is nearer to death than life. ([1929] 2006: 95)

The capacity to decide in any decisionist account leads to the thirst for a decision. The primary result of this is a need to act. When power is this poised to do so, time and again, it becomes the *normal* it so wishes to avoid, becoming routinized and so very much *predictable.* More troubling for a vitalist ideology that seeks control and independence, the result is quite the opposite—complete dependence on the need to *prove* that it deserves the aesthetic plateau it so desires. It becomes devoid, in the words of Wolin, from any "intrinsic content" in lieu of the "need of an external pretext of 'occasion' to realize itself" (1992: 443). A secondary result of this thirst is that the execution of the act itself must be done *quickly,* thus the emphasis on *doing something* rather than acting "cor-

---

42. From this discussion, we should not be surprised that recent work has delineated the intersection between Arendt and Foucault regarding power and the constitution or production of the subject, despite their seemingly incongruous "philosophical and metaphysical" perspectives (Allen 2002).

43. For the wider differences between Arendt and modern neoconservatism, see Owens 2007. It should be noted that Foucault also thought that instead of speaking about "an essential freedom, it would be better to speak of an 'agonism'—... less of a face-to-face confrontation which paralyzes both sides than a permanent provocation" ("The Subject and Power," in Foucault 1982: 222).

rectly" to attend to the problem. Here again aesthetics are involved, as the decision maker must fall back on a vivid memory or even image during a crisis when time is of the essence, as psychological accounts of decision-making suggest (Vertzberger 1990: 62).

IV. The Generation of Aesthetic Power and Conflict

Having attempted to establish that the Self of power is generated aesthetically and that vitalism contains intrinsic vulnerabilities, I now outline, in this section, how each generation of a body of power engages, "cleans," and regenerates those aesthetic images. I suggest here not only that generations (in this case, of the nation-state) stimulate the aesthetic vision of rhythmic, psychological, and imaginative power but that their efficacy in creating this power itself depends on a sense of aesthetic legitimacy, even beauty, in order to bring about a shift. It begins with Tocqueville's observation that "among democratic nations, each generation is a new people" (1958: 2.105). Thus, while we might find a nation-state constrained by a history it may never escape, generational analysis instead sees them as key to understanding how the vanguard of youth engages a political community as if it were a "frontier."

Generations are key to understanding how power is generated—how it is aesthetically productive. They therefore help explicate the imprint that counterpower can make—ideologically, materially, ontologically, and aesthetically. The transformation of a national setting via generations is comprehensive. It is, of course, not just ideological or psychological; it is material as well. As each generation replaces an old one, it occupies higher positions in the economy, and it uses this wealth to inform institutions and renovate their purposes.[44] They begin to influence popular culture, the media, and the arts of a society. Such constructs even impact the style and cadence of speeches. These generational differences regarding the aesthetic Self are further exacerbated by technological developments (photography, Internet, wireless, blogs, video, etc.), which deliver what Foucault might title new "technologies of the Self." In terms of rhetorical content, emerging generations reconstruct the auto-

---

44. Bacevich (2005: 81) writes about second-generation neoconservatives, whose task "of spelling out an 'imperial self-definition'" precipitated the need to create new and unique institutions, such as the journal founded in 1995 by William Kristol, the *Weekly Standard*.

biographical narratives of "the idea that is" their nation-state[45] and furthermore erect the "regimes of truth" that demarcate the boundaries of discursive authority. Generational shifts also are important, for they signify a revision of a particular collective self-narrative that is invested with all types of aesthetic components—including symbols or tropes of common formative *experiences*—which are in part a rejection of a previous generation's disastrous practices that defile the community Self. Yet while submissions of a collective Self of power is "passed down," this sensory basis of the experience is not. Thus, generations of Americans sharing intragenerational experiences engage the subjectivity of the U.S. Self in a varied fashion over time. This can, furthermore, manifest into intergenerational rivalries over the Self—to the point where interpretive agreement over national history and identity is nearly impossible. Further, in their aesthetic production of this Self, generations carry with them, like vitalist ideologies, pockets of traumatic experiences that are vulnerable to counterpower manipulations. My account draws on several bodies of generational or "domain" analysis from a rather eclectic mix of scholars—including Thomas Kuhn's account of the rise and fall of scientific paradigms, Michael Roskin's (1974) depiction of foreign policy paradigms, and some of Morgenthau's work.

Kuhn's decades-old seminal deployment of "paradigms" is well known, but the role of paradigms in terms of mediating images and the aesthetic basis for the rise of emerging paradigms have gone largely unnoticed. Paradigms are a social phenomena. While they provide the conceptual tools for emerging students of science to use within the community they enter, the paradigm must be learned—acquired through "membership" (Kuhn 1962: 11). They are a form of communal discipline, "science as a social process" (Buger and Gadinger 2007: 96). But they do not just demarcate the boundaries of knowledge production. Paradigms are "a prerequisite to perception itself. What a man sees depends upon what he looks at and also upon what his previous visual-conceptual experience has taught him to see" (Kuhn 1962: 113). In Kuhn's account, experience shapes the ability of the scientist to see.

---

45. Yet by including generational interventions over these narratives, I differ slightly from the view that this narrative contains a linear logic, as the narratives constructed in the wake of traumatic events—discussed by Edkins, for example (2002: 248)—have usually been understood. Instead, the narrative here is cyclical or overlappingly parabolic, albeit that the incoming generation must adeptly cast their narratives as connected across time and thus present them as linear.

Most interesting is Kuhn's description of the important paradigmatic transitions in his portrayal of "scientific revolutions." Kuhn posits that all paradigms face counterinstances, even anomalies, but not until a crisis emerges in the scientific community. A new paradigm arises to confront this crisis when a "gestalt switch" (Kuhn 1962: 111–15) occurs, so that "after a revolution scientists are responding to a different world" (111).[46]

Aesthetic elements play a role in Kuhn's analysis, indicating more than just "counterinstances" that help propel the gestalt shift. During these shifts, these transitions, the aesthetic vision of a paradigm must be reinscribed. After listing "all [of] the arguments for a new paradigm . . . based upon the competitors' comparative ability to solve problems," Kuhn explains that "another sort of considerations" for rejecting the old paradigm in favor of a new one are

> the arguments, rarely made entirely explicit, that appeal to the individual's sense of the appropriate or the aesthetic—the new theory is said to be "neater," "more suitable," or "simpler" than the old . . . The importance of aesthetic considerations can sometimes be decisive. Though they often attract only a few scientists to a new theory, it is upon those few that its ultimate triumph may depend. (1962: 155–66)[47]

What Kuhn contributes to the current thesis is the pathbreaking notion that science does not just progress linearly and that the social and demographic construction of scientific paradigms leads to conflict—that there are emotional and aesthetic, rather than "rational," arguments that produce these shifts.[48] Two points are worth mentioning here in terms of Kuhn's analysis. First, if this aesthetic and emotional process occurs in the hardened, testable terrain of physical science, then it is an

---

46. Morgenthau paralleled this assessment, noting that the "departure from tradition evokes resistance from those who are intellectually . . . identified with that tradition . . . It is a burden to have to admit that what one and one's contempories had believed to be the truth was actually error" ([1964] 1970: 226).

47. Schmitt, in his 1929 essay "The Age of Neutralizations and Depoliticizations," depicted a similar function for the aesthetic—a domain that provided a via media between "the moralism of the eighteenth [century] and the economism of the nineteenth century" ([1929] 2006: 84).

48. In this respect, Morgenthau saw science as a "cultural endeavor," thereby making it indistinguishable from such fields as religion, philosophy, literature, and art ([1964] 1970: 226).

even more likely attribute in the process of the construction of collective Selves (national paradigms collectively engaged by citizens). This is, in many ways, the point Roland Bleiker was driving toward when he suggested that the line between mimetic and aesthetic representations was quite blurred, in that the former relies in part on the latter in the form of a "romantic ideal [in] the autonomy of the Self, the quest for independence and self-determination, [and] the belief that people can shape history" (2001: 516).

In this, Kuhn finds what some IR scholars might find to be a surprising ally—Kenneth Waltz. Waltz noted, in the very first chapter of *Theory of International Politics,* that theories are made "creatively" and that "the longest process of painful trial and error will not lead to the construction of a theory unless at some point a *brilliant intuition flashes, a creative idea emerges*" (1979: 9, emphasis added). Thus, even in what many consider to be a founding canon of "rationalist" IR theory, we find such spontaneity and creativity.

We also gain a second insight from Kuhn in how the production of knowledge is a disciplinary, emotionalized process, invested in part with aesthetics and power. Therefore, as the deployment of "academic-intellectual parrhesia" in chapter 3 indicates, knowledge production itself is a cycle of power that is subject to intense eruptions through counterpower force.

Michael Roskin (1974) applied Kuhn's framework to the foreign policy decision-making process, asserting that generations engage common experiences in paradigmatic fashion to influence the method through which they approach the world once they get into power.[49] Roskin notes one "important provision" in applying Kuhn's model to foreign policy paradigms: that "because neither [of two competing paradigms] can be objectively verified in an indeterminate world," a "new foreign policy paradigm is merely *internalized*" (1974: 566). Yet, as in Dewey's discussion of the ruptured aesthetic experience and Kuhn's thesis on scientific revolutions, a "strong emotional component" (ibid., 565) accompanies intergenerational foreign policy paradigm conflicts. So strong is this emotional component, and "because they are antithetical, compromise [between two generations] is impossible. The two generations with their

---

49. Frank Klingberg first posited the oscillation of American "moods" between interventionism and introversion (isolationism) in his 1952 article "The Historical Alternation of Moods in American Foreign Policy" in *World Politics.*

different assumptions *talk past each other*" (ibid., 567, emphasis added).[50]
For Roskin (567), it is not only, however, generational changes ("the new
paradigm wins because it gains more younger adherents, while the advo-
cates of the old paradigm retire and die off") that fuels this emotionally
charged transition. It is also the observation that "catastrophes" occur—
shocks or ruptures—that are interpreted by the incoming generation to
have been a result of the old paradigm's mistaken assumptions. The old
paradigm is stale, "stuck in the past," and no longer capable of refresh-
ingly engaging the here and now.[51]

Not all generations are "alike" in the sense that the only thing that
separates them is the presence of intergenerational experience. This is
implied in Roskin's account, of course, but there is a sort of "cognitive
mapping" that recent generational analyses have conducted regarding
their cyclical proclivities, at least as they have evolved in the United
States. The typology was first proffered by Strauss and Howe (1991), pre-
senting four "types" of generations that recycle every two hundred
years—idealist, reactive, civic, and adaptive. The updated account of this
theory defines idealist generations as being "reared in an indulgent manner and . . . driven throughout their lives by their deeply held values." In
the United States, this includes the Baby Boom Generation of George W.
Bush and Bill Clinton. Reactive generations "tend to become alienated,
risk-taking, [but also] entrepreneurial and pragmatic in adulthood."
Generation X, represented to some degree by Barack Obama (who was
born in the "fault line" between the two generations, in 1961), is a reactive generation. Civic-minded generations, represented by the so-called
Millennial Generation in the United States and, in a previous cycle, by
the GI (or Greatest) Generation, "are reared in a highly protected manner so that an orientation to societal challenges, problem solving, and institution building marks their adult lives." Adaptive generations, such as
the Silent (or Forgotten) Generation in the United States (born in

50. This leads us to an interesting juxtaposition with the conclusions drawn from Maja
Zehfuss's (2007) fascinating aesthetic investigation of the *Wounds of Memory*. A key insight
of her study is that the processes of remembering and forgetting such wounds are not op-
posites but in an "inextricable relationship" with one another (71). Generational analyses
such as Roskin's suggest that the wounds of memory, whether we wish them to or not,
"heal" on their own as new formative experiences and reactions to those replace old ones.
We can not remember what we do not experience—we can not forget it either.

51. Jervis's well-known first hypothesis on misperception—"decision-makers tend to fit
incoming information into their existing theories and images" (1968: 455)—can be extrapolated here to an entire generation, so that over time, an increasing amount of information is transfixed solely through that generation's existing images and theories, and that
generation finds itself an increasing distance from the *social* and *societal* realities of the day.

1925–45), are, according to these authors, "risk averse, conformist, and inclined toward compromise" (Winograd and Hais 2008: 25–26). While such a typology may simplistically totalize generations, it is helpful to the extent that it indexes the general manner in which each type of generation both imbues purpose to national power and paints the image of the state with that power, within the confines of common experience.

Strauss and Howe's typology is most fascinating, and while it contains certain drawbacks (and what I describe later in this section as an analytical "trap"), it still provides several implications for the national subjectivity of power. Idealist generations in particular are most remarkable for the themes of this chapter, for two reasons. First, they are the ones who tend to be the most "vitalist" and concerned with the appearance, diet, and maintenance of the human body, and they seem, therefore, most prone to visual disgust at certain occasions and euphoric stimulation on others. For instance, Abraham Lincoln's Transcendental Generation, which occupied U.S. institutions from the early 1840s until a "realigning" election of 1868, defined beauty according to the lithe and thin body form. This idealization contrasted with celebration of the more rotund body form by the incoming, reactive Gilded Generation (Strauss and Howe 1991: 210). The Missionary Generation, emerging onto the American politico-cultural scene in the early 1900s and remaining in positions of authority through member Franklin Roosevelt's presidency (1945), emphasized new forms of dieting, as Upton Sinclair's *The Jungle* revealed the disdain and specific disgust the generation held over the processing of meat (Strauss and Howe 1991: 239).

Second, idealist generations, while challenging institutions and traditions in emerging years, aver toward intense moralism as they capture societal institutions. Reference here the loyalty oaths and sedition acts pushed through during World War I, the Missionary Generation's support of Prohibition during the 1920s, or the rise of the Moral Majority and the intense absolutism of the neoconservative boomers. A third and final observation is that while generational analysis intuits aesthetic shifts occurring rather infrequently—dependent on intergenerational conflict, displacement, and occupation—idealist generations are the most likely to "fragment." Thus, they are the most volatile sites for intragenerational disagreement. According to Strauss and Howe (1991), this occurred most violently with the Transcendental Generation (during the U.S. Civil War), but it also happened, to a lesser extent, during the Missionary Generation's reign (evidenced by the movie *Birth of a Nation*) and, of course, most recently with the Baby Boom Generation and its fragmentation over cultural issues and, most strikingly, over the Vietnam War it fought in. While several studies accord with Neta Crawford's observation that "the Vietnam War evokes powerful emotional associations for a generation of US soldiers and citizens . . . [which have made] the lessons of that war a significant factor in post-Vietnam foreign and military policy" (2000: 142 n. 95), the discussion of Vietnam that unfolds in chapter 5 of this book indicates that while baby boomer Americans collectively engage Vietnam as a counteraesthetic experience, as a "hole" in the idea that is the U.S. Self, they have nevertheless taken different lessons from that experience in terms of what type of challenge it provided to the U.S. Self.[52]

This is another aspect of Campbell's (1998) analysis that can be extended. Campbell asserts that "as an imagined community, the identity of a state is the effect of ritualized performances and formalized practices that operate in its name or in the service of its ideals" (130). Such rituals are vitally necessary for America, which is, according to Campbell, "de-historicized," which gives American "history the quality of an eternal present" (132). In a slightly different reading, as it is an aesthetic creation, the Self of power is formed in reaction to common experiences.

Paradigms have historically been characterized as influencing the manner in which people process information—to the point where such production of knowledge is not only not questioned but internalized. Kuhn calls the calm epochs periods of "normal science," where concepts are naturalized. During shifts, the fault lines of paradigms reveal emotional fissures. Thus, "the acrimony accompanying foreign policy paradigm shifts . . . suggests a strong emotional component" (Roskin 1974: 566). Because they share intragenerational experiences and use those experiences to understand what is the nation, different experiences across generations lead to emotional and even combative disagreements about the national interest.[53] According to Crawford, "an individual or generation that has firsthand or bystander experience with a highly emo-

---

52. Bacevich calls this collective group the "forces of reaction," in terms of revising military institutions in the wake of Vietnam (2005: 6).

53. Roskin includes a fascinating vignette detailing the transition from the isolationist "Versailles" generation in the wake of World War I to the more interventionist Pearl Harbor generation: "The 'isolationists' fought the growing interventionism every inch of the way. Strong emotions came to the surface. 'I could scarcely proceed further without losing my self-control,' wrote Secretary of State Cordell Hull of a 1939 confrontation with Senator Borah in which the latter disparaged State Department cables on an impending war in Europe. Other sources said that *Hull actually wept* at the meeting. It took the catastrophe at Pearl Harbor to squelch the obdurate bearers of the Versailles paradigm" (1974: 577–78).

tionally charged event will likely have strong memories of that event" (2000: 141).

Like all cyclical theories and accounts (such as classical realism, for example), generational analysis may at first blush appear to be somewhat "fatalistic" or tragic, a new generation stepping over the ashes left by the calamity of the previous one(s). Indeed, it feels like that sometimes, and this generational analysis is to some degree a "closed" perspective that, like in Kuhn's account, is parabolic rather than linear. Yet we return to a central intersection between generational analysis and counterpower: the moment of these shifts or breaks may be patterned and, in retrospect, seem like an obvious outcome of a stubborn generational cohort's mode of thinking, but because of the spontaneity and the totality with which they surround the community, they defy preparation. Like the "crisis" for Kuhn, the foreign policy disaster in Roskin, what Morgenthau once termed the "historical moment" (in his analysis of Spain's generational experience of 1898),[54] and the decisive moment in Schmitt,[55] they focus on an iconic "event" that "triggers" a shift in "the nation's mood" (Winograd and Hais 2008: 192).[56] Strauss and Howe title these "social moments"—so named because of the collective experience shared by that generation. Tellingly, Winograd and Hais use the word *traumatic* to explain these events (they provide Hurricane Katrina and the Iraq War as examples [193]), and such a word is perhaps appropriate considering the intersection of this analysis with studies on trauma, war, and memory (Edkins 2002, 2003).

Further, technological innovations not only provide "new media" for the emerging generation; they also facilitate two important processes (Winograd and Hais 2008: 24). When demographically substantive enough, a generation entering early adulthood (the eighteen to thirty-four demographic) becomes the primary target for the marketing of new

---

54. Morgenthau notes that the Spanish's loss in the war against the United States "threw the Spaniards back upon themselves, [but perhaps also] bec[a]me a necessary condition of national regeneration" and that "what moved the Spaniards so deeply and drove them into despair of their destiny was the manner in which this war was begun, carried on, and lost" (Morgenthau [1959] 1962: 214).

55. "It is a sudden rupture, a sudden temporal discontinuity creating a shock which produces horror" (Huysmans 1998b: 582; see also Wolin 1992: 433).

56. Such generational analyses have an analytical kinship with cycle theories of U.S. history (Klingberg 1952; Schlesinger 1949; see also Huntington 1991). American political studies known as "realignment theory," while controversial and a source of scholarly debate (see Mayhew 2004), have been used to better understand those U.S. elections where gigantic electoral shifts occurred, as well as for general understandings of levels of voter participation (Burnham 1970).

technologies, "gadgets" or products. Members of that generation become comfortable with these technologies so that they can use those to dislodge, disturb, and then remake existing community power structures. They thus serve as a medium, a "harness" for a generation to define their country's ontology—"who it was as a nation" (ibid., 50).

Of course, there are two "traps" to utilizing generational analyses, which I must disclose. First, age is a continuous variable. Thus, while the importance of the formative and traumatic moment or crisis impresses a national group, the manner in which this impacts a nation may be more or less uniform, across so-called generations. Second, as Strauss and Howe (1991: 36–37) readily admit, their generational framework may only be applicable in the case of the United States. They note, via Tocqueville, that the egalitarian nature of the United States forces generations to "remake" the ontology of the nation, and these accounts paper over the intragenerational conflicts that may arise and that indeed have arisen in the case of the Baby Boomer Generation. Such conflict is not over what specific "formative experience" defines that generation. Rather, the positions are over what to take from that traumatic rupture or experience.

On this point, however, there is some room for those who are optimistic regarding the presence of transnational generational processes. For starters, while its content does not entirely translate in a one-to-one categorization of such cohorts to modern generational analysis, generational identity, according to Laura Nash, did obtain in ancient Greece, "for the ages have a distinctive character and set associations" (1978: 5). More recent analysis by Ulrich Beck and Elisabeth Beck-Gernsheim entertains the notion that one can escape the "trap" of what they term "methodological nationalism" of generational analysis by explicating an "intertwined transnational generational" set of "constellations" (2008: 1). The common formative experience derived from globalization and the global market suggests that younger workers contain a mode of solidarity that exists transnationally (8–9). Most recently, in a manner that speaks to the role new technologies play in generational change, social commentators like Andrew Sullivan have likened the recent upheavals in Iranian politics to the influence of the Millennial Generation, apparently the same one that "elected Obama in America and may oust Ahmadinejad in Iran." He explains, "They want freedom; they are sick of lies; they enjoy life and know hope . . . This generation will not bypass existing institutions and methods: look at the record turnout in Iran and the massive mobilization of the young and minority vote in the US. But they will use technology to displace old modes and orders."[57] The Iranian experience, which remains, as of this writing, rather fluid, demonstrates the perpendicular sociological vectors that shape incoming cohorts—the national and the generational. Iranian Millennials are not as emotionally connected to the Iranian revolution of 1979 and thus are willing to oppose those in power who were responsible for that revolt against the Shah. Yet they are also deploying their protest through an Islamic and national context, using slogans—such as "Allah O Akbar"—originating from that 1979 revolution.

Conclusions

This chapter advanced the position that power can be understood as an aesthetic subjectivity that itself is explicated through three strata—the psychological, imaginative, and rhythmic—which work on one another in its construction and, conversely, in its destruction. The importance of the aesthetic and its relationship to power was demonstrated by reference, in part, to the philosophies of John Dewey and Michel Foucault as well as various aesthetic IR studies. This was used to suggest, however, that power is vulnerable to intense ruptures because such aesthetic subjectivity idealizes the Self of a community. Two further processes of self-creation fuel this regime of self-generated vulnerability. The first of these is the fetishization of demonstrated physique and movement through a decisive action as celebrated in vitalism. The second involves the processes by which generations form. Generations bring particular schema to bear on both the production of knowledge and a communal Self. These vulnerabilities help facilitate the impact of counterpower practices, three forms of which are explored in the following chapters, beginning with the practice of reflexive discourse.

---

57. Andrew Sullivan, "The Revolution Will be Twittered," *Atlantic Monthly*, 13 June 2009, http://andrewsullivan.theatlantic.com/the_daily_dish/2009/06/the-revolution-will-be-twittered-1.html.

TWO

Reflexive Discourse and Flattery as Counterpower

At a news conference on 27 December 2004, a bleary-eyed Jan Egeland, the UN undersecretary-general for humanitarian affairs, provided the latest updates regarding the massive humanitarian crisis unfolding immediately after an earthquake in the Indian Ocean had precipitated a series of tsunamis that impacted bordered areas. Asked a general question about the practice of foreign aid, Egeland responded,

> We were more generous when we were less rich, many of the rich countries. And it is beyond me why are we so stingy, really, when we are—and even Christmas time should remind many Western countries at least how rich we have become. And if actually the foreign assistance of many countries now is 0.1 or 0.2 percent of their gross national income, I think that is stingy, really. I don't think that is very generous. (United Nations 2004)

The U.S. response to these comments was remarkable. As the "we" indicates, Egeland was referring to all Western countries and their general aid contributions, yet for many Americans (and their leaders), it seemed that Egeland was specifically targeting the United States.

We would probably, at first glance, consider this to be an overly sensitive American response. Yet by the following day, the Bush administration would increase the amount pledged to tsunami relief to $35 million (from the initial $7–15 million), and by the end of the week, the pledge amount increased again to $350 million. Meanwhile, Bush officials indignantly defended America's foreign aid donations. In this sense, the "rational" material needs of a state seemed less important than disconfirming the *words* an official working for an international organization used to describe the United States. This was shocking behavior for an administration and a broader "exceptionalist culture" that generally bemoans the corruption of the United Nations and its own commitments to foreign aid donation in general.

What mattered in the aftermath of the Asian Tsunami was not only the U.S. commitment to the international community but its commitments and responsibilities to how it viewed itself. Being considered "stingy" challenged the aesthetic vision of a pristine U.S. Self. As a result of his general remarks on foreign aid, Egeland triggered a contentious debate in the United States about who or what America was, and it also generated the basis for the need to react to this aesthetic challenge.

This chapter investigates and develops two levels of what I call *reflexive discourse,* the dialogue of one actor to insecuritize a materially powerful target into acting according to the latter's sense of aesthetic integrity and self-identity. These two levels are influenced, in turn, by two accounts of social theory. Level 1 explicates how the ontological security of a powerful actor is challenged by reflexive discourse and that this can be explicated using some insights from Giddensian sociology. Level 2 explicates how reflexive discourse includes elements of what Foucault titles "flattery"—tactics I discuss as "bundling" and "self-flagellation." While Egeland used the latter two elements to help assuage the damaged ego of the United States (smoothing out the aesthetic rupture of his "stingy" remark) and thus helped generate the aesthetic context for American action thereafter, flattery also can be what Foucault calls a "mendacious discourse" which *artificially* inflates the "Self" of the target. After developing the concept of flattery, I provide a brief illustration of its "mendacity" in the form of the dialogue that the Iraqi National Congress, a dissident group living in the United States prior to the Iraq War, used to convince the United States to intervene on its behalf.

The chapter begins by situating reflexive discourse among other approaches to discourse (including post-structuralist, communicative, and instrumental approaches) and then examines the U.S. response to the recent Asian Tsunami and how the UN undersecretary-general for humanitarian affairs (inadvertently) used reflexive discourse immediately following the disaster when he suggested that Western nations were being "stingy" with their initial aid offers. As the work of Patrick Jackson (2006) and others (see Kratochwil 2007) advised, I do not make the case that Egeland *intentionally* used reflexive discourse; it is more likely that he accidentally stumbled upon, with the American response, a process that is not usually assumed to operate in the policy world.[1] I focus here on the context in which this discursive counterpower event materialized and the effects or empirical patterns that emerged thereafter. I do not analyze, however, whether the U.S. reaction actually increased the welfare of tsunami victims. Such an outcome, while ethically important to determine in other respects, is beside the point in an account of counterpower. Here, I instead find it necessary to explicate how discourse such as Egeland's can generate the reaction it did from the United States.

Several aspects of this particular case illustration are intriguing. First is the context in which it occurred: an intransigent American administration, recently empowered by reelection and historically skeptical and even hostile to international institutions and especially the United Nations, was nevertheless so concerned with the words used by a UN official to, in its view, describe the United States that it increased by twenty times the amount of pledged aid to affected areas of the tsunami. This indicates what was mentioned in the previous chapter regarding the "narcissism" of aesthetic power. Second, and related to this, is how the reaction forced the Bush administration not only to react rapidly but to also try to reestablish the rhythm of a U.S. autobiographical narrative that was disturbed by Egeland's (general) criticism. Let us begin to resolve this puzzle with a form of counterpower that explicates the manipulative abilities of an asymmetric discourse.

Reflexive Discourse

The reflexive discourse approach is used in this chapter to understand how centralized bodies of power can be stimulated into action. This is one of the conclusions that I made in a study that reviewed the impact of Abraham Lincoln's Emancipation Proclamation on Britain's ultimately neutral stance toward the American Civil War.

> The larger issue is that discursive representations can be just as powerful as physical presentations of force—because they can compel other international actors to "do what they otherwise would not do." The possibility is that states not only know what actions will make other states *physically* insecure, but also *ontologically* insecure as well. (Steele 2005: 539)

---

1. I discuss the issue of Egeland's intentions later in this chapter.

I argued in that article how the proclamation changed the meaning of the American Civil War for Great Britain and thus changed what a British intervention would have meant to British self-identity. In this chapter, I assert, in like fashion, that the same self-identity process that the proclamation engendered obtained, ever briefly, in the United States following Egeland's "stingy" comment. Such comments play on the insecurities these bodies of power—including powerful nation-states—have about what their policies (on aid or on inaction over genocide, for instance) mean or have meant for their sense of self-identity.[2] The reflexive discourse approach can be understood through two levels.

Level 1: Self-Identity and the Biographical Narrative

Self-identity tells agents what they want in international politics and how they should go about getting it. As Rom Harre notes, the self "is not an entity. Rather it is a site, a site from which a person perceives the world and a place from which to act" (1998: 3). Reflexivity is defined by Giddens as "the monitored character of the ongoing flow of social life" (1984: 3).[3] This monitoring relates to the (auto)biographical narrative that agents set up to explain their actions, and thus reflexive monitoring is implicated in the process of self-identity.[4] This narrative represents the self-monitoring of actions for agents, and it also publicizes commitments to the international community.[5] The biographical narrative represents the best approximation of what a state's actions mean to its sense of national "Self," and it is integral to the securing of self-identity through time.

2. Giddens (1991) terms self-identity "both robust and fragile. Fragile, because the biography the individual reflexively holds in mind is only one 'story' among many other potential stories that could be told about her development as a self" (55). Reflexive discourse uses this malleability to stimulate a targeted state into acting according to a different "story line."

3. Reflexive discourse thus asks targeted states to engage in a "reflexive response," whereby such states, like the individual, are "asked by another to supply 'a reason' or 'reason' for, or otherwise to explicate, certain features of his or her activity" (Giddens 1984: 73).

4. For more on reflexivity, see Guzzini 2000. Reflexive monitoring does not necessarily lead to a progressively "better" subject, as noted by Petr Drulak: "Contingency works either way and social innovations can be both good and bad" (2006: 143). McSweeney (1999: 140) differentiates reflexivity and "self-reflection."

5. The fact that states justify their actions (based on some type of "principle" or "norm") distinguishes, for Hedley Bull (1977: esp. 45) and many English School theorists, an international society from an international system, since justifications for even transgressions will lead to more "international order."

To recall the discussion from the previous chapter, narratives satisfy all three strata of the aesthetic subjectivity of power. Their routinizing function is rhythmic; they create ambiguous but idealized images for the individual citizen to engage in an imaginative way; and, finally, they help provide for those citizens a psychologically satisfying locus of the Self by scaling out the chaos of other, potentially endless Selves. This locus is, of course, transformed by each generation that captures the institutions of a state—as these dominate the biographical narrative state agents provide to the international community and themselves to justify their actions. Such movements craft language, providing both a cadence and a style, as well as a discursive substance, that reflect their conception of their nation-state's sense of "Self."

This "locating" function thus enlarges the minimal (but important) responsibility that some realist authors have assumed for the state agent. Take, for instance, the position by George Kennan: "Government is an agent, not a principal. Its primary obligation is to the *interests* of the national society it represents, not to the moral impulses that individual elements of that society may experience . . . The interests of the national society for which government has to concern itself are basically those of its military security, the integrity of its political life and the well-being of its people. These needs have no moral quality" (1985/86: 206, emphasis added). In crafting a biographical narrative, however, state agents do more than ensure the "integrity of political life" by *locating* a national Self. State agents are thus "embodied speakers" for the state, not only "enabl[ing] an understanding of the location of the embodied speaker, but also [as] a means through which responsibility is taken for being positioned," and thus there exists "an ethical dimension to selfhood" (May 2000: 164).

While it by no means homogenizes a state's self-identity—indeed, the securing of self-identity is a complex and highly political process—the biographical narrative scales down the relevant roles a state occupies for particular sets of situations. In the words of Jutta Weldes, "*in order for the state to act,* state officials must produce representations" of a crisis, a crisis that itself "depends on the discursively constituted identity of the state" (1999: 57–58, emphasis added). The narrative itself is an outcome of intrasocietal debates, but because it emanates from the agents of states, it nevertheless most closely approximates the identity commitments a state will pursue in international relations.

Giddens (1991: 54) notes how identity is "not to be found in behavior, nor—important though this is—in the reactions of others, but in a capacity *to keep a particular narrative going.*" Yet the content of behavior is important to that narrative, as the latter "cannot be wholly fictive. It must continually integrate events which occur in the external world and sort them into the ongoing 'story' about the self." It is this narrative that counterpower engages and disturbs. The narrative invests the Self with an aesthetic content that is challenged through reflexive discourse.

Level 2: Flattery and Self

This is perhaps where Giddens's assumptions about the Self need to be amended, in several ways. First, a drawback to solely utilizing a Giddensian model of the Self's construction of identity in counterpower's discursive terms derives from the tendency of such models to focus, to varying degrees, on continuities and routines, rather than the "disruptions" that occur from what Giddens titles "critical situations" (1984: 10–11). In a counterpower account, the analytical focus is the unpredictability (the "critical situation," if you will), rather than the continuity of routinized behavior. Reflexive discourse as a form of counterpower must be formalized or structured enough regarding the Self of the target to be recognized but must be flexible enough to be unpredictable, lest it get swamped and categorized by a re-formed power. What makes reflexive discourse a form of counterpower is this flexibility—its spontaneity, as it catches a target "off guard," disrupting the cadence and structure of a narrative with reference to inconsistencies.

In a related second way, the discursive manipulation of ontological security has been largely overlooked in its mainstream IR treatments, which assume that "ontological security seeking" links to public forums—where apparently innocuous "public commitments" could "foster habits of reflection" (Mitzen 2006: 363). This move attempts to link the Giddensian Self to Habermasian IR approaches via the mechanism of "mutual recognition." While making somewhat intuitive sense, it operates largely on the assumption that there is a welcoming "public" that might force an ontological security-seeking agent in-line with its community precepts. The problem with this view, as well as with much Habermasian-influenced IR theory, is that it does not recognize the implicit forms of power and manipulation of that power that can occupy these public spaces. For instance, return to Giddens's view that the autobiographical narrative cannot be "wholly fictive." While the autobiographical narrative of the Self cannot be wholly fictive per se, other actors might interdict and manipulate that narrative with their own forms of "events," themselves incomplete or inaccurate. Those events may then be integrated by the target into an autobiographical narrative and an inflated sense of Self, transforming its logic for action. It is therefore more vulnerable, being pushed away from its original moorings, as this smoothed sense of Self becomes the basis to act in an external world, a "reality" that is likely to be misunderstood by the Self. To understand this form of manipulation, we can again turn to Foucault, specifically his account of flattery.

A flatterer can be what Foucault titles a "depraved orator" who "only say[s] what the people want to hear" (2001: 82),[6] yet flattery also allows "the inferior to win over the greater power he comes up against." While the flatterer "reinforces" the superior's superiority, he can also "succeed in diverting the superior's power" through a "mendacious discourse" where the "superior will see himself with more qualities, strength and power than he possesses" (Foucault 2005: 375–76). Flattery can thus render the target "impotent and blind." How is this discourse a form of counterpower? For one, it "fills in" the holes of insecurity that consume the target—the aesthetic insecurity discussed earlier in this book. Yet this filler is hardly genuine, the flatterer "speaks, and it is by speaking that the inferior, boosting the superior's extra power as it were, *can get what he wants from him*" (375, emphasis added). The superior power, more superior now than ever before, becomes dependent on this "foreign discourse" (378), making the superior "think that he is the most handsome, the wealthiest, the most powerful, etcetera" (375). Second, flattery is a form of counterpower precisely *because* the powerful are narcissistic; they are consumed by either "disgust" or "self-love," which "leads to being attracted by sensual pleasures" (378), and thus the powerful target becomes *dependent* on this "lying" to maintain its false sense of strength and power. So flattery can make the targeted superior much more vulnerable than it would have been otherwise.

Perhaps a brief empirical illustration is appropriate here, as I think we can reconstruct as a form of flattery the discourse used by dissident groups or "exiles" to influence their host countries' policies through persuasion. Take the recent example of the Iraqi National Congress (INC)—a group of Iraqi dissidents founded in 1992 with the express purpose of toppling Saddam Hussein's regime. Living in the United States and the United Kingdom, this group helped provide various

---

6. I distinguish flattery from the parrhesiastic "truth-telling" form of counterpower in chapter 3.

amounts of information and intelligence to these countries in the run-up to the Iraq War. Some of the information provided by one dissident, Adnan Ihsan Saeed al-Haideri, "found its way into the [U.S.] National Intelligence estimate on nuclear weapons in October 2002 as the Bush administration prepared its case for war" (Roston 2008). This information claimed that over three hundred secret weapons facilities in an Iraqi program of weapons of mass destruction had been "reactivated" during the run-up to the war (Ricks 2006: 56).

Other members, such as the leader of the INC, Ahmed Chalabi, were especially adept at coaxing interlocutors with flattery.[7] In one account, titled tellingly "The Manipulator," Chalabi is said to pride "himself on his understanding of the United States and its history" (Mayer 2004). Another example comes from a November 2002 panel discussion regarding a possible U.S. intervention into Iraq, where yet another Iraqi dissident, Kanan Makiya, used flattery with reference to particular themes that resonated with his "liberal" American audience. This account was detailed in a *New York Times* article appearing in December of that same year. Such perceived knowledge of a target's history is important for what I call the "bundling" tactic of flattery, which I discuss in more detail later in this chapter.

Yet unlike Foucault, we should not assume that a flatterer's cajoling against power will ultimately result in "wickedness." It could (and arguably did in the case of the INC and the inaccurate intelligence it provided the United States), but as the Egeland example discussed in this chapter seeks to establish, while flattery inflates a target's aesthetic power and makes for a "harder fall" when this subjectivity of power is ruptured, it (flattery) can also help smooth out the original shock to aesthetic self-identity and lay a foundation (albeit a semiartificial one) from which power can re-act. In the case of the U.S. reaction to Egeland and Egeland's bundling strategies that occurred thereafter, the result was a more forceful aid response by the United States.

Approaches to Discourse in IR Theory

We can situate reflexive and flattery discourse alongside three existing approaches to discourse that permeate IR theory: strategic or instru-

---

7. One account notes how Chalabi was able to convince "the Kurdish leader Jalal Talabani that eating ice cream would help him lose weight and was good for him . . . Ahmed's lie in that case was innocuous. *He had a knack for telling people what they wanted to hear:* about ice cream, democracy, and Iraq" (Roston 2008: 225–26).

mental approaches; communicative, argumentative, or discourse ethics approaches; and post-structuralist approaches (see table 1). Strategic action is defined by Habermas as "one actor [seeking] to *influence* the behavior of another by means of the threat of sanctions or the prospect of gratification in order to *cause* the interaction to continue as the first actor desires," whereas in "communicative action one actor seeks *rationally* to *motivate* another by relying on the illocutionary binding/bonding effect of the offer contained in his speech act" (1990: 58). When actors treat each other strategically, the "coordination of the subjects' action depends on the extent to which their egocentric utility calculations mesh" (135).

When applied in IR theory, strategic approaches to discourse (see esp. Elster 1996) find a theoretical niche in the neorealist and neoliberal accounts and, ironically, share certain assumptions with post-structuralist analyses as well (Campbell 1992; Doty 1996). While post-struc-

TABLE 1. Approaches to Discourse

Form	Strategic (instrumental)	Post-structural	Reflexive	Communicative (dialogic)
Assumptions	*Coercive Speech* • Speech is backed by force/sanctions • References to power • Serves material interests	*Concealment Speech* • Masks power relations	*Identity Speech* • Power relations assumed • Less powerful "stimulate" powerful by engendering aesthetic insecurity	*Ideal Speech* • Reference to Archimedean or constructively shared "morality" • Collective morality and collective identity commitments
Core referent	Self-help (egoism)		Ontological security/ Aesthetic power/self-identity	Intersubjective understandings; "common lifeworld"
Power referent (Barnett and Duval 2005)	Compulsory	Productive	"Reverse" compulsory and productive	Deemphasized contaminant
Equivalent IR perspective	Neo-neo synthesis	"Superrealist" post-structuralism	Critical ontological security	Constructivist/Mainstream ontological security/ English school/ Relationalism

turalists obviously have important critiques of mainstream approaches to IR, they all assume that power relations consume the ether of international politics. For Mervyn Frost, postmodern analysts are actually "superrealists," in that while

> they reject the realist commitment to the *state,* they remain firmly committed to the realist canon that the primary focus in all social analysis must be on *power* . . . Throughout, the aim is on bringing to light *structures of power* which were previously hidden by, for example, the silences created by other theories. (1996: 69, emphasis added)

According to Frost, both groups of theorists see the most powerful agents engaging in instrumental discourse, while both the form their power takes (explicit vs. implicit) and the method in which it is exercised vary.

Yet post-structuralist approaches uncover important and unique processes that distinguish them from strategic approaches. To begin, strategic approaches contain a "compulsory" logic of power, where an actor utilizes as a background condition its resources in discourse "to advance its interest in direct opposition to the interests of another" actor (Barnett and Duval 2005: 50). This overt assumption of the nature of power in strategic approaches is a bit naive, according to the post-structuralist. If power were so inherently overt, it would meet much more resistance. Instead, since power relations remain largely obscured, such power must be excavated from the many subversive processes and practices that structuralist accounts consider to be unimportant.[8] Edkins explains, "Discourse is always a question of the production or legitimation of power, and all power (though inevitable and necessary) is ultimately illegitimate. Discourse serves to conceal, to cover up, that illegitimacy" (1999: 59). The post-structuralist locates power in the medium of *language,* which sets the boundaries for reality. In Barnett and Duvall's taxonomy of power, "productive power" accords to this view of the post-structuralist, where such power refers "to the discursive production of the subjects, the fixing of meanings and the terms of action" (2005: 56).

Reflexive and flattery discourse share some terrain with strategic approaches. Both accept that targeted audiences are self-interested and

---

8. "Power is seen by the realists to work on the surface. For postmodernists, however, power operates beneath the surface. It is a web that permeates all relationships, institutions, and bodies of knowledge" (Sterling-Folker and Shinko 2005: 637).

(albeit to varying degrees) ego-driven, and flattery discourse especially recognizes the notion in strategic discourse called, in the words of Habermas, "prospects of gratification." The issue of power asymmetries is not only assumed in reflexive and flattery discourses; it is necessary (as I discuss in more detail later in this chapter). Yet there is one vitally important distinction: the direction in which the primary influence takes place (a less-powerful speaker stimulating a powerful actor) is exactly the opposite of the assumed direction in instrumental discourse; thus, reflexive and flattery discourses entail a "reverse" compulsory logic.

More recent work on verbal influence strategies comes from Janice Bially Mattern's "postconstructivist" conceptualization of "representational force." Such a conceptualization is quite similar to reflexive discourse.

> Representational force is an expression of force like any other: It is a form of power wielded in a blunt manner so as to radically limit the options of the subjects at whom it is directed. But rather than being mediated through material or physical expressions, representational force is wielded through the structure of the language that an author chooses to use in forming representations. (Bially Mattern 2005: 95)

The purpose of representational force, like reflexive discourse, is to "trap" an intended target into an action that will serve identity drives: "representational force . . . traps victims with the credible threat emanating from potential violence to their *subjectivity*" (96). Yet while she asserts that this subjectivity is tied to the "encompassing sense of Self," Bially Mattern locates the sense of Self through a different process.

> In this way, subjectivity is intersubjectively constituted through shared self-other relationships and the sociolinguistic constructs through which *they* are constituted . . . It is made "real" by the social (and so linguistic) relationship through which I have come to know my self in the world in relation to others. (ibid.)

Thus, while reflexive discourse shares a theoretical affinity with "representational force," it focuses, as a form of counterpower, almost exclusively on the subject's aesthetic insecurity as a basis for action. This is also the foundation for distinguishing reflexive from communicative approaches.

Reflexive versus Communicative Discourse

One can characterize reflexive discourse's relationship with communicative approaches in two ways. The more moderate position is that reflexive discourse may not be in tension with communicative discourse but, rather, occupies a "middle" position between overt coercion, on the one hand, and consensus (through argumentation), on the other (Steele 2007: 909). From this position, there are some important assumptions held in common by reflexive discourse and forms of communicative approaches—the latter being represented within IR theory through the work of scholars such as Thomas Risse and Neta Crawford (see also Payne 2001). First, it may be the case that reflexive discourse, as a form of counterpower, occurs within regimes of communicatively inspired societal transformation. Here, then, the major difference that remains between reflexive and communicative discourse is temporal, but both could be occurring simultaneously. Argumentation transforms, in a transcendental sense, collective identity over much longer periods—for instance, Neta Crawford's (2002) discourse ethics argument's timeline is five hundred years—whereas reflexive discourse, as a form of transgressional counterpower, can stimulate power to act in a matter of weeks or months.[9]

A second point of intersection is that reflexive discourse is a form of "ethical argument" most closely resembling "identity arguments" that

> posit that people of a certain kind act or don't act in certain ways and the audience of the argument either positively or negatively identifies with the people in question . . . Identity arguments work by producing or calling upon previously existing identities and differences among groups and claiming that specific behaviors are associated with certain identities. (Crawford 2002: 24–25)[10]

Crawford asserts that ethical arguments worked to promote decolonization because

> early advocates of colonial reform and later proponents of decolonization called on colonizers to act in ways that were consistent with

---

9. However, it should be noted that for Jackson (2003), such a rhetorical common place "needs to be activated through the articulation of specific arguments drawing this implication from the commonplace; on its own, a commonplace—even a 'civilizational one'—does nothing" (239).

10. See also Cruz 2000.

their (evolving) identities . . . Ethical arguments about reforming
colonialism harnessed the emotions of embarrassment and shame on
the part of colonizers. (2002: 388)

In a moderate sense, this is what reflexive discourse can produce as
well—by referencing the self-identity of a targeted state, the speaker asks
a targeted audience to justify the disconnect between its narrative understanding of self-identity and the seemingly contradictory actions or
inactions.

Therefore, both reflexive and communicative actions might be forms
of verbal persuasion. In the specific realm of humanitarian action, Risse,
Ropp, and Sikkink (1998) use a communicative approach to identify the
"shaming" process nongovernmental organizations have used, referring
to this process as "moral persuasion," which is "mainly *about identity politics,* that is, Western governments and their societies are reminded of
their own values as liberal democracies and the need to act upon them
in their foreign policies" (251, emphasis added). These scholars acknowledge one problem, however, for communicative action: "this
process of moral consciousness-raising can take quite a long time"
(ibid.).

To see why this is so, we must consider that reflexive discourse, first,
does not assume that the discourse must *change* the identity of a targeted
power (although, as stated already in this chapter, identities do vary over
time and space). At any one time or even over periods of time, centralized bodies of power do not recognize any large discrepancy between
their actions and their professed identities. This is not always intentional—as individuals, we perform certain actions, out of habit, that we
do not recognize are completely inconsistent with how we identify ourselves. Once these actions are brought to our attention *by others,* not only
might we seek to rectify the disconnect, but we feel ashamed to have performed those actions in the first place. The problem that reflexive discourse assumes (and what makes it necessary) occurs when no actors call
out this disconnect. Since the routinized habit (in Giddensian terms) of
"contra-identity" actions actually shield a centralized body of power from
recognizing any incongruity, such exposure is necessary. But, again, it
does not or need not result in the changing of a targeted power's identity, it must simply force it to reflexively engage that sense of identity in
light of the biographical narrative and the conditions that threaten its
sense of Self.

Yet the moderate position—that reflexive and communicative approaches are only distinguished on the basis of the rate of stimulation—ends here. The more assertive position that I wish to advance is that communicative approaches have several other problematic assumptions about power, collective "universal morality," and thick intersubjectivity that inhibit their analytical ability to explicate counterpower events. First and foremost, communicative perspectives assume a "common lifeworld": "a supply of collective interpretations of the world and of themselves, as provided by language, a common history, or culture. A common lifeworld consists of a shared culture, a common system of norms and rules perceived as legitimate, and the social identity of actors being capable of communicating and acting" (Risse 2000: 10). As Krebs and Jackson aver, for those approaches, "politics is less about contest than consensus, less about powering than puzzling, and deliberative exchange consequently takes center stage. Ideally, actors leave power and rank at the door, and they seek to persuade others and are themselves open to persuasion" (2007: 39).[11] Once power is set aside, speakers approach an "ideal-speech situation" as coequals. This common lifeworld produces, and is reproduced by, collective streams of identity, as Elinor Ostrom states: "Exchanging mutual commitment, increasing trust, creating and reinforcing norms, and developing a group identity appear to be the most important processes that make communication efficacious" (1998: 7).

Admittedly, there is a wide range of IR accounts that have loosely approached a "common lifeworld" from a variety of angles. For instance, Ronald Krebs and Patrick Jackson (2007), while staying away from the methodologically problematic position that one can "know" what actors are thinking or intending by what they say (42), still maintain that their identified analytic of rhetorical coercion is "premised on a political community that shares at least some understandings of the boundaries of acceptable discourse. The more tightly linked the community, the greater the possibilities for rhetorical coercion" (55). Yet if a community is so tightly linked, what, then, is the need for "coercion," rhetorical or otherwise? Needless to say, such an assumption of the common lifeworld is more limited in reflexive discourse. While communicative theorists like

---

11. Krebs and Jackson (2007: 39) cite the following constructivist studies as following Habermasian communicative action, although "sometimes only implicitly": Johnson 1993; Müller 2001; Risse 2000. I believe a case can be made that the following works also reside within this framework: Cruz 2000; Crawford 2002.

Risse admit that "the degree to which a common lifeworld exists in international relations varies considerably according to world regions and issue-areas" (2000: 16), this somewhat obscures the length of time required to develop a collective lifeworld.

Why else is the lifeworld assumption problematic? For one, it may have been constructed as an ideal model of communication where "a debating and coffee drinking public depended upon a distinctive configuration of liberal social and economic interests and could not, even according to Habermas, feasibly (or desirably) be transposed to late-modern society" (Owens 2005: 44). It should be noted that Habermas himself has jettisoned it as an assumption from his social theory anyway, as it "wrongly suggested that the way in which normativity is embedded in communicative practice corresponds to the Kantian notion of a regulative idea." Further, the lifeworld can be "semantically colonized" by "steering media" such as power and money, an area that constrains communicative freedom instead of making it possible in the first place (Herborth 2008: 25–30). Nevertheless, this has opened Habermas to critics who contend that "he is blind to the 'power-practices' that occur" in the lifeworld (Kogler 1996: 27; see also Thompson 1999: 202).

In a similar fashion, Patrick Jackson has noted that Habermasian approaches in IR implicitly link the nature of the arguments that "win out" to some "transcendental validity" (2006: 22). Thus, Jackson's position (2003, 2006), while discussing "rhetorical commonplaces" where a sense of community is "nested" (2003: 239), areas that are "topological resources" for actors to use to persuade others, nevertheless sees these commonplaces as "only weakly shared" (ibid.). This is precisely one distinction that must be made in this more aggressive position that I am advancing between reflexive discourse and communicative action.[12]

The issue of the lifeworld is also important in the context of an unfolding crisis. Neta Crawford made a valiant attempt to develop a discourse ethics approach to humanitarian intervention (HI), and accord-

---

12. The distinction between Jackson's noted work (2003, 2006) and counterpower mostly results from his emphasis on "relationalism," which locates "causal mechanisms in the *intersubjective* realm surrounding and penetrating actors. Relationalists . . . argue that properties of *context* rather than properties of the *individual*, should be the analytical focus" (2003: 234). The location of the target is, in his accounts, "in the intersubjective space *between*" two interlocutors (2006: 32), whereas the force of counterpower is produced within the targeted power itself, not in the "context" between a counterpower agent and a targeted power.

ing to her account, the complex problems of HI are a result of the lack of definitional *consensus* (2002: 426). The purpose for dialoguing the issue is not to objectify HI so much as it is to *intersubjectify* the conceptual issues of HI, because

> without a conversation open to all where all presuppositions and arguments are open to challenge, humanitarian intervention may become a practice that resembles colonial interventions . . . Without a wider and sustained conversation, truly humanitarian interventions . . . may become more rare. (432)

But the problem (which Crawford seems to recognize)[13] is that consensus building around the highly charged political issue of HI—a practice that enforces one contested "principle" (human rights) of international society at the expense of another (sovereignty)—is an elusive goal.[14]

On this issue, a contrary view is that some of the most materially capable states who can best ensure "humanitarian" outcomes are not going to engage in humanitarian interventions out of a sense of responsibility to "international principles" that might require them to act. This is not to say that Crawford's proposal on this issue lacks merit—she lists some important benefits from a running dialogue on HI, and if there is to be a "long-term" solution to HI, it must be constructed from the conversation that Crawford proposes. Thus, we can consider yet again the manner in which transgressional counterpower (in this case, reflexive discourse) exists within a transitional or transcendental process (in this case, communicative discourse ethics).

Returning to the literature on HI, such approaches assume that because HI results in the intended rescue of *others,* the actions must be a function of a state or group of intervening states feeling responsive to an international ethic or "norm." As an example, Nicholas Wheeler asserts that states derive their motives for humanitarian action in rules.

---

13. "A convention would take years to accomplish, and no doubt more than one humanitarian crisis will arise before a convention is achieved" (Crawford 2002: 432). Crawford also adds that this should not preclude the construction of such a convention, since there are also benefits to such engagement as well (433).

14. For attempts at engaging a pluralist conception of human rights, see both Booth's and Brown's contributions to the Dunne and Wheeler (1999) volume *Human Rights in Global Politics.* One possible (and partial) solution to Crawford's dilemma is that such dialogue has been present for some time. Wheeler's seminal book demonstrates that states have at least recognized an informal right to circumvent sovereignty to uphold a minimal solidarist form of humanitarian rescue.

> [Rules] are guidance devices that tell us how to act in particular circumstances and that proscribe certain forms of conduct as unacceptable . . . Norms are not material barriers and their constraining power derives from the *social disapproval* that breaking them entails. (2000: 4–5, emphasis added)

Two problems with this conception emerge. The post-structuralist might note how social disapproval itself is a form of discipline, even if it is communal in nature, and that the rules for legitimation themselves stack the deck for the powerful. To wit, Owens mentions that Wheeler's Habermasian-inspired portions of work distinguish between power as "domination and force" and "power that is legitimate because it is predicated on shared norms" (Wheeler 2000: 2), but as Owens notes, Wheeler "does this without considering whether the 'norms' themselves may actually be domination and force" (2006: 48). Indeed, this is one broader problem with deliberative models of legitimation.

The works of both Morgenthau and Foucault, which foreground power, productively extend this critique. Morgenthau focused on the limits of human beings and advised that

> while [one] is making use of suasion he must not be oblivious to the role of power, and vice versa, and of each he must have just the right quantity and quality, neither too much nor too little, neither too early nor too late, neither too strong nor too weak. ([1952] 1962: 331)

So for Morgenthau, human beings can be persuaded, but only if power is acknowledged and if the complexities of all human relations are made central. Additionally, he focuses on the timing of such suasion—"neither too early nor too late"—and further asks "how much suasion and power is present on the other side at a particular *moment* of history, and how much and what kind is likely to be present tomorrow?" (ibid.). This advises the speaker to bear in mind the history of the targeted actor and their *development* ("likely to be present tomorrow") during a particular series of exchanges.

For Foucault, it is less the moral code or "norm" that does the work of persuasion and more the *relationship to one-self,* which thereafter does the "work" of stimulating action. Reflexive discourse thus focuses on the targeted state's capabilities to effect external change as a part of this aesthetic subjectivity of power. Subjects may refer to universals (i.e., defending "Western civilization") as a mode of their own subjectivity,[15] but to understand what they are doing after being stimulated, we need an ontology of the present, the historical present.

Therefore, reflexive discourse jettisons the requirement in some communicative approaches to remove references to power. Furthermore, in reflexive discourse, there is no reason for actors to "recognize each other as equals" as there is in communicative discourse (Risse 1999: 534). Instead, power asymmetries are explicit and public, assumed by all and utilized to focus action. Because power asymmetries are explicit, a speaker who uses reflexive discourse acknowledges the limits of its own agency and, conversely, the increased ability a targeted state has to produce an intended outcome.[16] Because a crisis requires quick attention and because powerful subjects can most capably attend to that crisis—indeed, may stake their aesthetic subjectivity on it—both the speaker and the targeted subject use that asymmetry as a core referent. The more this referent is acknowledged, the better.

Flattery: Massaging the Target through Bundling and Self-Flagellation

There are important caveats regarding who reflexive discourse should target in attempting to stimulate a powerful state to act. First, a state's self-identity is *politically negotiated* because the "stakes" of self-identity are important (determining "who we are" influences "what we need to do"). State agents obviously play an important role in this negotiation for political reasons. Second, while reflexive discourse engages the *state* rather than state agents, which seemingly makes the criticisms less personal, such criticisms will still, inevitably, draw the reactionary ire of the targeted state's population. Even though the discourse is directed at no one individual, because of the aesthetic strata of power and (as Dewey showed) the public role of the citizen in crafting it, reflexive discourse is

---

15. This may be why communicative approaches, like that of Risse, Ropp, and Sikkink (1998), have a confusing understanding of "shame." Contrary to their view that nongovernmental organizations (or other international actors) can "shame" states into acting according to *community* understandings of human rights, shame itself is understood as a private sense of transgression, a disconnect from the Self rather than the public codes of the community.

16. While they can not be entertained in this book, there are some directional affinities between reflexive discourse and the ideology of "revolutionism" identified by Martin Wight (1991) and impressively developed in the context of colonial encounters by Tim Dunne (1997).

still "personal." Thus, speakers using reflexive discourse may use two tactics that attend to this concern. The first of these, which I call "bundling," involves packaging the self-identity disconnect criticisms (which reference a current unaddressed situation) with praise for past situations in which the targeted state genuinely lived up to its self-identity commitments. Bundled praise can be general or specific ("in World War II, you liberated Europe and defended millions from tyranny"). This locates past situations that are most likely a targeted state's source of pride, and it can be used to "massage" state agents into acting again.

For a brief empirical illustration of bundling, we might return again to the Iraqi dissident example. In addition to the false information that his organization provided U.S. intelligence agencies regarding Iraqi programs of weapons of mass destruction, INC leader Chalabi also constructed a vision of what a post-invasion Iraq would "look like," using, as his model, past instances from U.S. foreign policy history that were a source of U.S. pride. The term *de-Baathification* itself was constructed out of the cloth of U.S. "de-Nazification" plans in post–World War II Germany (Roston 2008: 237).[17] According to one scholar, the use of the word *de-Baathification* "was in line with the INC's practice of invoking the lessons of America's occupation" of both that country and Japan. Further, while

> the Nazi analogy was historically flawed . . . it proved an effective public relations gambit. It is important that [for Americans] World War II had none of the moral baggage, or the associations with quagmires, that plagued discussion of the war in the Philippines in 1898, the occupation of Panama in 1989, the Vietnam War, the invasions in the Dominican Republic, or elsewhere. Invoking the World War II parallel not only provided moral clarity but put to rest worries about what would happen after the invasion, which was one of the major concerns being raised at the time by those opposed to the war. And it worked brilliantly. (ibid., 238–39)

Another example of flattery comes from a November 2002 panel discussion at New York University dealing with a potential U.S. intervention in Iraq. As reported in a *New York Times* article titled "The Liberal Quandary over Iraq," most of the participants at the panel were liberal intellectuals, who, for the majority of the discussion, detailed "reasonable ar-

---

17. Saddam Hussein's regime in Iraq was part of the Ba'ath Party, which had maintained control of Iraq periodically since the early 1960s.

guments against a war in Iraq." An Iraqi dissident named Kanan Makiya, also a panelist, spoke last. George Packer, a liberal advocate (then and now) of intervention, poignantly and dramatically relayed the scene.

> He said: "I'm afraid I'm going to strike a discordant note." He pointed out that Iraqis, who will pay the highest price in the event of an invasion, "overwhelmingly want this war." He outlined a vision of postwar Iraq as a secular democracy with equal rights for all of its citizens. This vision would be new to the Arab world. "It can be encouraged, or it can be crushed just like that. But think about what you're doing if you crush it." Makiya's voice rose as he came to an end. "I rest my moral case on the following: if there's a sliver of a chance of it happening, a 5 to 10 percent chance, you have a moral obligation, I say, to do it."

The impact of the statement was "electrifying. The room, which just minutes earlier had settled into a sober and comfortable rejection of war, exploded in applause. The other panelists looked startled, and their reasonable arguments suddenly lay deflated on the table before them." The effect was, in Packer's view, "because Makiya had spoken the language beloved by liberal hawks . . . He had given the people in the room an *image of their own ideals*" (Packer 2002, emphasis added).[18]

Thus, what Chalabi and other INC members were doing was not just "seducing" the United States through analogies (Khong 1992) but further conjuring an aesthetic vision of a past "painting" that was itself in part constructed by the United States or at least a sympathetic intelligentsia—the vision of two democracies "nation-built" out of previous totalitarian dictatorships. They used this vision as a salve to soothe the United States into acting.

A second tactic that a reflexive speaker might use would be *self-flagellation*. A typical reaction by those facing criticism is to turn the focus back on the speaker. In this case, targeted states might ask what the speaker's organization is doing (or has ever done) to address the crisis. By recognizing their own organizations' (state or nonstate) shortcomings, a speaker admits that they are incapable of confronting a situation on their own. Because the power asymmetry is an open fact, a speaker can

---

18. On the eve of the Iraq War, when momentum for the war was probably unstoppable, the liberal interventionist Packer began to be somewhat cautious in the "lofty ideas" of Makiya (Packer 2003). I discuss the liberal-intellectual infatuation with "progress" and the ignorance of power in the following chapter regarding academic-intellectual parrhesia.

use this to justify why they cannot deal with the crisis and to assert that the powerful state is the only one that can. In the event that the speaker's credibility, rather than just capabilities, is put on trial by a targeted state, self-flagellation would require the speaker to admit to some credibility gaps by engaging in self-critique, without completely acquiescing to the argument of the targeted state.

To Be or Not to Be "Stingy": American Self-Identity and the Asian Tsunami

On 26 December 2004, an earthquake measuring over 9.0 on the Richter scale erupted off the west coast of Sumatra in the Indian Ocean. The resulting tsunami spread destruction along coastal areas of several countries, including Indonesia, India, Sri Lanka, Somalia, Tanzania, Malaysia, Maldives, and Kenya. The most recent estimates for the dead and missing put the toll around 230,000 (UN tsunami envoy). Needless to say, for those who did not perish, the devastation created a massive humanitarian crisis, and most crucial to survival were necessary amounts of food and water and combating the spread of disease. Serving as the organization most responsible for coordinating the international community's response, the UN lobbied for a robust and quick outpouring of aid and materiel from its member states.

A reflexive discourse approach provides insight into why comments made by UN undersecretary-general for humanitarian affairs Jan Egeland immediately following the Asian Tsunami created such a backlash in the United States. At a 27 December press conference, Egeland issued a request by noting the urgency with which aid was needed.

> We are appealing now for all donor nations to respond generously and open their coffers now at the end of the year. We know some donors now have some money still available on this year's budget and we need them now. (United Nations 2004)

Egeland was shortly thereafter asked whether aid "dealing with the tsunami may eclipse or even undercut, both in terms of money and personnel, U.N. relief efforts elsewhere in the world, such as Sudan." He responded,

> It is really a problem that for too many rich countries, the pie is finite. You take out a slice and there is less for the rest. And I think an un-

precedented disaster like this one should lead to unprecedented generosity from countries that should be new and additional funds. Some others have the same sum for all disasters in the world, and I'm afraid for the coming year because there are several donors who are actually less generous than before in a growing world economy. (ibid.)

Another questioner asked, "When you were talking about donor countries in a growing economy giving less, are you prepared to name them?" As mentioned at the beginning of this chapter, Egeland responded,

We were more generous when we were less rich, many of the rich countries. And it is beyond me why are we so stingy, really, when we are—and even Christmas time should remind many Western countries at least how rich we have become. And if actually the foreign assistance of many countries now is 0.1 or 0.2 percent of their gross national income, I think that is stingy, really. I don't think that is very generous. (ibid.)

This statement is noteworthy to consider from the angle of aesthetic power in two respects. First, Egeland did not specifically single out the United States (a later comment by Egeland that I discuss shortly referred to the "populations" of the United States, the European Union, and Norway, Egeland's native country, but collectively, rather than individually). He referred to "rich countries" and "Western countries" when generalizing. Second, Egeland was not considering the specific response to the Asian Tsunami but was instead generalizing about the generosity of donor countries as measured by their foreign assistance as a percentage of gross domestic product. As his comments just quoted indicate and as his statements in the days that followed those comments substantiate, there is no evidence that Egeland intended or targeted these comments to offend any Western country, let alone the United States, although Egeland was probably attempting to urgently stimulate an increase in aid, as his comment "we need them now," about donor nations, suggests.

But the response to these comments in the United States was remarkable. Like a coiled spring ready to take offense, many perceived the comments to target not just "the West" but more specifically the United States. By the following day, the Bush administration had increased the pledged amount from the United States to thirty-five million dollars (from an initial seven to fifteen million), all while Bush officials defended America's foreign aid donations. In the opening remarks to a press gaggle on 28 December, the deputy White House press secretary stated, without provocation,

> And I would just say, of course, that the United States and the American people are the single largest contributors to international aid efforts across the globe. We have been for the past few years and I have every expectation that that will continue. (Duffy 2004)

U.S. secretary of state Colin Powell appeared on various morning shows that day and also defended American foreign aid amounts: "The United States has given more aid in the last four years than any other nation or combination of nations in the world" (U.S. Department of State 2004c). Other Bush administration officials qualified the initial pledge amount of fifteen million dollars as reflecting a wait-and-see approach to the crisis: "I think it's important to note that everything I've just described in terms of assistance is preliminary . . . We know the needs will be greater. This is a disaster of almost unimaginable dimension and it's going to require a massive support for some time" (U.S. Department of State 2004b).

A 28 December article in the neoconservative daily *Washington Times,* inaccurately but tellingly headlined "U.N. Official Slams U.S. as 'Stingy' over Aid," began with a description that positioned the initial U.S. donation as the target for Egeland's comments.

> The Bush administration yesterday pledged $15 million to Asian nations hit by a tsunami that has killed more than 22,500 people, although the United Nations' humanitarian-aid chief called the donation "stingy." (Sammon 2004)

The neoconservative Heritage Foundation, in an article posted 30 December titled "American Generosity Is Underappreciated," reported that Egeland had

> criticized the U.S. commitment as "stingy" despite the fact that the U.S. pledge far exceeded those of all European nations. He quickly apologized and said that he did not mean to single out the United States, *but the transcript of his comments clearly identifies the U.S. as the primary target.* (Heritage Foundation 2004, emphasis added)

Of course, the transcript of Egeland's comments indicates no such identification of the United States, but of importance here is the reaction by a neoconservative think tank.

We should sit back for a moment and therefore recognize how striking this reaction was. These were words used by an official of an organization that neoconservatives universally disdain—an organization considered to be feckless and unproductive (Bolton 2008). That these conservative American publications reported Egeland's comments as targeting the United States is probably more important, then, than even the reality of Egeland's true intentions (or even the general comment Egeland issued, which did not mention the United States once).

Yet in another respect, the reaction made logical sense, for as a vitalist perspective on power, action, and movement (as Podhoretz's previously referenced outlook attests), neoconservatives hold that the United States needs to realize its sense of Self in a crisis. A shirking from acting against this comment would reveal a weakness of the U.S. Self. Thus, there was a need, in this case, to quickly cleanse U.S. power of this displeasing moniker of "stingy." By 31 December, the Bush administration increased the pledged amount again, this time from $35 million to $350 million. Bush's publicly stated reason for this tenfold increase seemed to have nothing to do with Egeland's comments and instead was the result of "initial findings of American assessment teams on the ground [that] indicate[d] the need for financial and other assistance will steadily increase in the days and weeks ahead" (Sanger and Hoge 2005).

Egeland's comments are an example of reflexive discourse, as they generated U.S. insecurity over America's actions compared to America's historical biographical narrative. The focus here should not be on intentions. There is no way of knowing, for instance, whether Egeland intended his comments to generate a reflexive debate within the United States or whether he knew that states have self-identity needs. In fact, both in public and in recent firsthand accounts, Egeland has said this was not his intention.

In his 27 December comments, Egeland also assumed that the populations of countries were driven by different calculations than their leaderships, that the citizens of these countries were motivated by something other than pure material interest.

And I have an additional point . . . Politicians do not understand their own populations, because all their populations in the United States, in the European Union, in Norway, which is number 1 in the world, we want to give more, as voters, as taxpayers. People say we should give what we give now, or more. Politicians believe that they are really burdening the taxpayers too much, and the taxpayers want to give less. It's not true. They want to give more. (2004)

Admittedly, not enough time has passed since the tsunami to know whether the administration's second (and more massive) increase in aid was due to Egeland's comments, an "on-the-ground" assessment, domestic political pressure, some combination of all of these, or something else entirely. Most probably, as I stated in the introduction to this chapter, Egeland inadvertently triggered a process that he did not expect or fully intend. This stated, four points follow regarding the reflexive process evident in this case.

First, as a reflexive discourse approach suggests, Egeland's remark "We know some donors now have some money still available on this year's budget and we need them now" indicates that the United Nations and aid agencies in general lacked the initial resources to confront the situation and needed help from the most materially capable states. The power asymmetry that is assumed in reflexive discourse was used by Egeland in his appeal, as an honest admission, an explicit admission, noting from where aid must materialize. Thus the element of power disparities, which would be reduced or "set aside" in communicative discourse, becomes the basis for influencing a targeted, "superior" actor.

Second, even though it was not the focus of Egeland's remarks, the perception within the United States was that its place and identity within international society was being challenged. Being considered "stingy" challenged not only the moral position of the United States within the international community but the aesthetic vision Americans had about the subjectivity of U.S. power, as many domestic opinion leaders and critics noted. For example, in a letter to Bush before the second increase in aid, Democratic senator Patrick Leahy specified, "I believe that the United States, with capabilities far exceeding those of other nations, has an obligation to play the lead role in responding to this humanitarian emergency" (Leahy 2004). Even after the tenfold increase of 31 December, Democratic senator Carl Levin remarked that he thought "we [now] appear to be on the right track," but he posed a critical stance toward the initial response and what it demonstrated about America.

But the first few days were disappointing, and not in keeping with *the great American tradition of generosity,* either in the size of that first announcement [of seven to fifteen million dollars], or the speed with which it was given. I thought the first few days of silence from the president was not in *keeping with that tradition.* (UN Aid Official 2005, emphasis added)

Furthermore, Egeland's comments retrieved, at minimum, enough insecurity within the Bush administration to compel agents to defend American policies regarding foreign aid. This is all the more remarkable considering that the subsequent challenges posed to the administration came from two resources for American ontological debates who had been largely ineffective since at least 9/11: the Democratic Party and the American media.

Third, Egeland's comments throughout the week indicate the bundling tactic described earlier in this chapter, a tactic that smoothes out the perforated edges of a rupture generated by a counterpower event. Besides the praise he issued regarding the desire of donor countries' populations to "give more," Egeland also mentioned, in an interview with American National Public Radio on 29 December,

Well, the US is a very, very generous contributor to our humanitarian relief; it's actually the biggest in absolute terms as a contributor . . . I did never, ever say that anybody did not respond well to the tsunami. (National Public Radio 2004)

Egeland stated in another interview, on the 1 January 2005 broadcast of *Fox News Sunday,* that he "did not want this kind of debate . . . I was confident, I was sure that the United States and the other of our partners would be generous in the wake of such an enormous natural disaster. And, as we have seen, I was right in that" (UN Aid Official 2005).

Fourth, it is fairly evident that Egeland's comments led to the massive increase in U.S. aid on 3 January. While the official U.S. line was that increased aid was due to new information from "assessment teams," one account in the *New York Times* stated: "As recently as Thursday [30 December], a senior State Department official deeply involved in the rescue efforts said Washington *had not received any word from any assessment team asking for more money*" (Sanger and Hoge 2005, emphasis added). The comment led, at minimum, to criticism of the Bush administration in the U.S. press, and this was not lost on administration officials. As recorded in James Traub's (2006) account of a private meeting at the United Nations that included Egeland, Powell, and Annan in attendance, after Egeland told Powell, "I'm sorry that I was misunderstood when I commented on the overall trend in the past year," Powell responded, "The more serious it gets, the more hysterical the press gets . . . They're looking for things to criticize us on" (278–79). Another fascinating exchange recorded by Traub reveals both Egeland's contrition and the contented observation by UN officials that Egeland's comments were indeed responsible for the increase in U.S. aid: "Diplomats and aid officials were thrilled to see a high-ranking UN figure stick his head so fearlessly in the lion's mouth; what's more, even as it growled, Washington made haste to up the ante . . . That morning [30 December], while several of us were in the waiting room outside Anna's office, Iqbal Riza [special adviser to the secretary-general] told Egeland, '"Stingy" has gotten their attention more than "illegal."' 'I wasn't talking about the tsunami response,' Egeland protested. 'I was just commenting on ODA [official development assistance].' 'You don't have to explain it to *us*,' Riza said with a smile" (276–77).

Thus, we might pose a counterfactual that asks what the increase would have looked like had Egeland *not* issued the "stingy" comments on 27 December and whether any domestic pressure coming from the U.S. media or the Democrats would have had the effect that it did (or whether these groups would have even felt compelled or empowered to issue any criticisms of the Bush administration's response in the first place).

The Limits of Reflexive and Flattery Discourse

To further develop the reflexive discourse approach, one might conclude that what is needed for reflexive discourse to consistently work would be a comprehensive "cognitive mapping" of each centralized body of power's aesthetic identity. While an exercise might provide important insights, it is not immediately required. International actors know that nation-states have certain sources of "shame" (at least in a metaphorical sense) and that these sources influence security interests—for example, that Germany is insecure over the Holocaust (Bach 1999) or that Britain exhibits anxiety over the "appeasement" of Hitler before World War II. As the work of Ayse Zarakol on Japan, Turkey, and ontological security suggests, states do engage these sources from time to time, although such sources are so touchy that state apologies for events such as the Armenian genocide (for Turkey) and World War II atrocities (for Japan) are rare due to the complexity of state ontological security (Zarakol 2010). It goes without saying, then, that international agents must avoid specific reference to these sources of shame while they attempt, at the same time, to get the targeted state to reflexively engage those sources (as holes in the aesthetic vision of the national Self) and to activate some aesthetic security response.

While I have discussed here one example of reflexive discourse during the Asian Tsunami crisis and provided illustrations of flattery, it should be noted that many national debates can be considered the reflexive engagement a society has with its own sense of self-identity. As I mention in chapter 4, the recent debate in the United States over torture as a policy in the War on Terror is sometimes framed with references not to American responsibilities to "international agreements" like the Geneva Accords but, rather, to America's responsibility to its founding "principles."[19] As a challenge to aesthetic integrity, linking reflexive discursive images to notions of U.S. "honor" and ideals can be very powerful.

Yet at the same time, as noted in chapter 1, the observation that power is partly psychological and gains its currency precisely because its aesthetic Self is "imagined" by collective bodies who participate in its perpetuation implies that there are limits to discursive forms of counterpower. Reflexive discourse may challenge these aesthetics but then soon face a disciplinary counterdiscourse. This is, I assert in chapter 4, what occurred in the case of Illinois senator Richard Durbin—who, in attempting to place U.S. ontology "on trial," likened post-9/11 interrogation techniques to those of Stalin's gulags or Nazi Germany's gestapo. While these comments provided a psychological challenge, they "overstepped" the bounds of decorum-based propriety, and Durbin was forced to apologize. But this occurred because words can be challenged by further words, whereas images are more immediate, more shocking. Thus, what I explicate in chapter 4 as "self-interrogative imaging" can provide a more potent form of counterpower, in certain cases, than reflexive discourse. Until then, we must turn to another form of counterpower, but one that in many ways dispenses with the primary discursive conventions that reflexive discourse uses to manipulate a target. Such notions of tact, resolution, massaging, or bundling are the nemesis to this form of "fearless speech," which seeks out conflict through the telling of truth. The parrhesia form of counterpower is explored in chapter 3.

---

19. One whistleblower during the torture debate, Navy general counsel Alberto J. Mora, remarked, "The Constitution recognizes that man has an inherent right, not bestowed by the state or laws, to personal dignity, including the right to be free of cruelty. It applies to all human beings, not just in America—even those designated as 'unlawful enemy combatants.' If you make this exception, the whole Constitution crumbles. It's a transformative issue" (Mayer 2006).

THREE

Truth and Power: Parrhesia as Counterpower

*The philosophers of friendship were, accordingly, concerned to discrimi-
nate the type of the flatterer or adulator from that of the true friend, and
the surest sign of the difference was candor and honesty—the parrhesia
characteristic of the true friend as opposed to the deceitfulness that
marked that parasite.* —KONSTAN 1996:10

LESTER: *Aw, man. You made friends with them. See, friendship is the
    booze they feed you. They want you to get drunk on feeling like you
    belong.*
WILLIAM: *Well, it was fun.*
LESTER: *They make you feel cool. And hey. I met you. You are not cool
    . . . My advice to you: I know you think those guys are your friends.
    You wanna be a true friend to them? Be honest, and unmerciful.*

—*ALMOST FAMOUS* 2000

What issues and difficulties arise and surround the telling of truth in international politics? An investigation into truth-telling as a form of counterpower must, of course, focus on the intersection between truth and
power, their relationships and tensions. Such an investigation also requires us to recognize how the practice of politics influences the extent
to which truth-telling disturbs the ontology of power, for, as Foucault
states, "the least glimmer of truth is conditioned by politics" (1984: 5).
Yet the purpose in exploring the effects of truth-telling is not to show
how a pristine, universal truth can itself be established and disseminated
in spite of power, nor is the purpose here to show how power can rein-

corporate a more just truth into its techniques of domination. Rather, I here argue that we might resist the temptation to establish solutions based on some idealized notion of truth, for such solutions, over the long term, will in any case be implemented by those with the resources to do so. Such solutions would then become techniques for further normalization, depoliticization, and domination. This investigation suggests that the proper place for truth is instead within and through, but not above, politics. A politicized truth creates possibilities.

Having discussed in the last chapter the practices of reflexive discourse and flattery, I here investigate a second form of counterpower, *parrhesia*—frankness, or the telling of truth. Like other forms of counterpower, parrhesia need not be part of an *intentional* strategy of resistance against power, but it can serve to strip away and reform it by manipulating the unpleasant edges of this power, forcing it to re-act. Section I of the chapter examines parrhesia's development primarily by Foucault (although a few other advances are noted), and while parrhesia refers to the "telling" of truth, this section articulates why parrhesia is less a form of discourse than it is a forceful moment or occasion. Section II discusses some general conditions that constrain truth-telling in politics. The chapter then deploys these insights through two exercises of modern-day parrhesia: in section III, Cynic parrhesia (as illustrated through its use by al-Qaeda); in section IV, academic-intellectual parrhesia, drawing on not only Foucault but also the work of Hans Morgenthau. Intellectual parrhesia is explicitly engaged through the case illustrations of the experiences of John Dewey during World War I and Morgenthau during the Vietnam War, as well as more contemporary examples stemming from the academy's posture during the recent U.S. War on Terror. These illustrations explicate the "seductive" qualities of power that serve to inhibit intellectual parrhesia.

I. Parrhesia

Foucault develops parrhesia in several ways, but it can begin to be understood as a verbal activity whereby the person using it is telling the truth. Foucault's most inclusive definition of parrhesia is "frankness, freedom and openness that leads one to say what one has to say, as one wishes to say it, when one wishes to say it, and in the form one thinks is necessary for saying it" (2005: 372). Moreover, it is "free speech, *released from the rules, freed from rhetorical procedures*" (406, emphasis added). Foucault investigated various forms of parrhesia in detail, most directly through two sets of lectures, those given in the spring of 1982 at the College of France (reproduced in *Hermeneutics of the Subject* [2005]) and in the fall of 1983 at University of California, Berkeley (reproduced in *Fearless Speech* [2001]). The *parrhesiastes* "says what is true because he *knows* that it *is* true, and he *knows* that it is true because it is really true" (Foucault 2001: 14).

Despite what Foucault might have us believe here, this is not Truth, in that what is true for the parrhesiastes is spatially and temporally conditioned. What is true, then? How do we, as analysts, know "the truth" when we see it? The short answer is that it represents what is truthful—what is uncomfortably so—*for power at that moment* in which counterpower engages it. It is the truth as the parrhesiastes senses it, against the power it is targeting. While this partially accords to Ken Booth's "view that there is no ultimate truth in the social world, only a pragmatic truth," such truth is *not* "created intersubjectively" (Booth 1999: 43) but, rather, constructed as a regime of self-knowledge in the subject being targeted, a subject whose Self is continuously created.[1] Temporally, this pragmatic truth is even more localized than this, as it is a micromoment within that continuous Self.[2] Put another way, we only know what we are right now or perhaps what we *just were*. This is why the concept of *kairos*, or the moment of parrhesia, is so integral to its counterpower possibilities, as I discuss later in this chapter.

Foucault distinguishes between two varieties of parrhesia: the form practiced in ancient Greece and the confessional form it took in early manifestations of Christianity. The differences, in turn, relate to unique views of "the Self." Conceiving of parrhesia as "frankness" relates, in one sense, to the "Christian ideal of perfect openness before God" (Konstan 1996: 15). Here, it is baring one's soul in front of others (being comfortable enough with friends to share everything) and before God, in that professing the truth of one's Self accords with this conceptualization of frankness. The parrhesia practiced in ancient Greece, however, did not archive life events or produce a narrative meant to draw a connection between the parrhesiastes and God as much as it was intended "to

---

1. In fact, this focus on self-creation in Foucault's later work has been said to characterize the Self as seeing the Other not as a fellow subject but, rather, as an "object" or "narcissistic extension of the primary subject" (McNay 1994: 153). Thus, the parrhesiastes is an extension of the insecuritized Self, rather than a subject through which intersubjectivity is constructed.

2. Bleiker (2000: 130) notes that "power relations . . . need to be assessed in their unique spatio-temporal setting."

show that there is a relation" between the parrhesiastes' words and deeds (Foucault 2001: 97–101). This is the form Foucault focuses on in his lectures and tends to see as more authentic and, I would argue, compelling in its ability to manipulate an interlocutor.

Parrhesiastic Force versus Communicative and Strategic Approaches

We can briefly address, as I did with reflexive discourse in the previous chapter, how parrhesia can be distinguished from communicative and strategic approaches to discourse. A first possibility is that parrhesia may not even be considered "discourse" if the latter is defined as stemming from "multiple sources that form a system of signification" (Sjostedt 2007: 237). It is perhaps more accurate to title parrhesia a dialectical force, one where, in the words of Adorno, "an opponent's strength is absorbed and turned against him . . . By means of logic, dialectics grasps the coercive character of logic" (1973: 406). Parrhesia can be continuous, but it can also be momentary, and it need not be communicative in the sense that the parrhesiastes hopes to acquire further information or knowledge from the interlocutor in another round of dialogue.

Like the reflexive discourse discussed in chapter 2, parrhesiastic force sees the Habermasian check of power and rank "at the door" to be very problematic. This is not because we ideally should not want to "wish away" power during modes of speaking but, rather, because, as a form of counterpower, parrhesia sees power inequity as *vital* to its forceful possibilities. This clarifies Foucault's critique of Habermas as "utopian" (1989: 446). Instead, the purpose should be not to "dissolve" this power but to "acquire . . . the practice of the self" that enables individuals to "play the game" of power (ibid.). Communicative discourse obscures the play of power, the specific counterpower possibilities that can be accessed by adept agents using parrhesia.

Moreover, the problem is not necessarily that power relations overwhelm discursive interaction, as the earlier work of Foucault (1972) might have us conclude. Indeed, if that were solely the case, then Krebs and Jackson's answer would be both necessary and sufficient: "it would . . . seem more helpful to adopt a theoretical framework that explicates the power of rhetoric even when politics is *not* truth-seeking and truth-generating" (Krebs and Jackson 2007: 40). The power of truth-telling need not turn into an attempt at building consensus, nor does it require power to be "set aside." Instead, we must recognize that resources for the countering of power lie within the interlocutor's sense of Self that is the target of parrhesia. Parrhesia becomes a "force" by turning the target "into a place where truth can appear and act as a real force through the presence of memory and the efficiency of discourse" (Foucault 2007: 164). Like reflexive discourse, what distinguishes parrhesia from strategic approaches to discourse is that, while power remains in the forefront of the speech,

> parrhesia is a form of criticism, either towards another or towards oneself, but *always in a situation where the speaker or confessor is in a position of inferiority to the interlocutor.* (Foucault 2001: 17–18, emphasis added)

In short, while parrhesia echoes the assumption, held by strategic approaches, that there is an asymmetric relationship (in terms of material power) between the speaker and interlocutor, the direction of manipulative relations is reversed. Similarly (as also discussed in chapter 5 of this book), Bleiker's account of transversal dissent views "discourses [as] not invincible monolithic forces" but, instead, "fragmented and thin at times," with possibilities for "dissent linger[ing] in these cracks" (2000: 206). The basis for parrhesiastic force is the self-knowledge of the target, its own ontology, which can be revealed and transformed during a truth-telling exercise.

Parrhesia versus Flattery

Despite some similarities just noted between the two, reflexive discourse, unlike parrhesia, assumes some interaction between participants. While it includes an unpredictable moment of counterpower that can puncture a target, reflexive discourse also includes further tactics of flattery intended to "massage" the target, what I termed "bundling" and "self-flagellation" in the previous chapter. More to the point, reflexive discourse is strategic and may take a bit of time to produce counterpower (although less time than the discourse ethics approaches analyzed in chapter 2). Because the "truth is suspended and frail, due to its temporal substance" (Adorno 1973: 34), parrhesia's influence is derived from its "momentary" ability to engage the powerful target. This is but one of the ways in which parrhesia can be distinguished from forms of "flattery." The condition of danger in which a parrhesiastes puts himself or herself results from the courage to "oppose the demos." "Depraved" orators, in comparison, "only say what the people desire to hear" (flattery) (Foucault 2001: 82). Thus, a further contrast between flattery and parrhesia is that one reinforces authority, while the other insecuritizes it.

As indicated in the quotations that opened this chapter, we also see the issue of friendship appear in discussions of parrhesia. Yet this sense of friendship is also, as the Foucault quote appearing in the introduction to this chapter indicates, "conditioned by politics." In democratic Athens, parrhesia could manifest itself within the web of friendship, but, again, this web was itself based on the "ideology of equality and *freedom from dependency* that was central to the Athenian civic ideal." This emphasis on self-reliance meant that while "friends were to be relied upon for assistance in times of crisis," all citizens had a duty to be frank to one another, as civic equals. Thus, parrhesia equated to the right of free, public speech (Konstan 1996: 8–9, emphasis added), a right that was important to Athenian democracy (Fredrickson 1996: 167). Yet when Athenian democracy fell, the idea of parrhesia was transformed within the political context where it operated, moving "to a concern with relations between people of unequal station and power" (Konstan 1996: 9).

Parrhesia is "antiflattery" in that its objective is to "act so that at a given moment the person to whom one is speaking" is no longer dependent on the speaker. The "truth" sets the target free, so to speak, in that it "guarantees the other's autonomy" (Foucault 2005: 379). Both Foucault and the foreign policy literature on groupthink focus on a major constraint, perhaps *the* major constraint, to truth in international politics: the social drive of conformity.[3] "Who will tell the Prince the truth?" Foucault asks. "Who will tell the Prince what he is, not as Emperor but as a man?" (2005: 381). I discuss this issue in more detail in section II, where I entertain other factors that inhibit "the truth" in politics—such as the need for politeness and "tact."

We might also ask whether Foucault is idealizing the notion of "truth" and its ability to manipulate power. This is possible, as Foucault tends to admire the Greeks as heroic exemplars, and he may be doing the same here with their version of truth-telling. Yet, at the same time, Foucault is not quite focusing on truth or truth-telling per se in these lectures as

3. First archived by Janis, groupthink is exemplified by instances of "mindless conformity and collective misjudgment of serious risks, [the latter of which are] collectively laughed off in a clubby atmosphere of relaxed conviviality" (Janis 1982: 3).

much as he is concerned with "how the will to truth produces effects of power" (Ashenden and Owen 2001: 9) and, I would add, countereffects *upon* power. The will to truth, rather than just truth, is the substance of parrhesia's counterpower possibilities. If Foucault is idealizing anything in his work, it is perhaps the notion that human societies, to varying degrees, all hold this "will" and that games of truth for human subjects are "obligatory."

Both parrhesia and flattery (such as reflexive discourse) are asymmetric challenges against power, but the distinction is the *direct* danger that truth-tellers put themselves in, for the fearlessness of the parrhesiastes a priori serves to disturb power. An interesting parallel regarding this asymmetry can be gleaned by comparison to what James Scott (1990) has termed the "art of resistance." Subordinate actors, he asserts, do not acquiesce to domination but resist through transcripts that are "hidden" from dominant actors. The "weak," Scott reasons, "have obvious and compelling reasons to seek refuge behind a mask when in the presence of power" (10). The dominant have an interest in determining what constitutes "publicly valid" transcripts. This seems reasonable, but parrhesia's force is derived from the subordinate's willingness to tell the truth within full earshot of the powerful. The problem occurs when such power is not challenged, when these transcripts *remain* hidden. This is not to argue, contra Scott, that a hidden transcript is not a form of resistance. Nor is it to disconfirm the notion that overt inaction by subordinates equates to acquiescence. Yet as a form of counterpower, parrhesiastic force's influence rests on its ability to reveal the contradictions of the targeted dominant actor. This is easier said than done, however, as the pitfalls faced by the parrhesiastes in international politics and its study are many, and it is to these that I now attend.

II. Concealing the Truth

What role does truth play in the political processes studied by IR scholars? Because the goal of identifying truth in politics remains elusive, several bodies of scholarship have examined the conditions that have constrained or served to conceal its presence. The voluminous literature on truth and reconciliation commissions (TRCs) has assessed their varying levels of success in transitioning societies toward some sense of order in a post-conflict era. How one concludes such studies depends on which societal value should be prioritized in these transitional societies—justice or reconciliation. For Amy Gutmann and Dennis Thompson (2000), social harmony should not be the goal of such commissions. Other studies, such as the work of Debra DeLaet, reveal how the gendered nature of norms embodied in TRCs served as obstacles to truth and "also limited the willingness of women to testify about themselves as victims" (2006: 168). Thus, two of the words in the name *truth and reconciliation commissions* exist in great tension.

The burgeoning work on "organized hypocrisy" in IR theory is also instructive here, as it collectively informs how an organization's environment constrains it from implementing policies, so that problematic outcomes arise "unintentionally" (Lipson 2007: 9). Hypocrisy is generated from "talk" that serves to "compensate for inconsistent action" (10; Brunsson 2002: xiv). Because it draws on sophisticated bodies of theoretical work on bureaucratization of organizations (Weber [1922] 1949), organizational environments, or cultures (DiMaggio and Powell 1983) and thus can utilize well-specified conceptual building blocks such as the "logics of appropriateness and consequences" (March and Olsen 1998), this work logically focuses on explaining the hypocritical behavior of international organizations (Weaver 2008; Barnett and Finnemore 2004; Bukovansky 2005). The inconsistency thus arises from conflicting values or interests that those organizations embody. These conflicts provide the pressures that lay the groundwork for hypocrisy.

The problem a parrhesiastes faces is more comprehensive than the organized hypocrisy of the power the parrhesiastes targets. Organized hypocrisy assumes that international organizations *know* that they are being hypocritical and just either cannot or will not change their actions to rectify this. The hypocritical agent is uncertain already, thus the need for the "cover-up" of "talk." But a parrhesiastes faces instead a situation where the audience does not even realize this inconsistency or at least does not recognize its own ability to change this behavior. This is a more stubborn beast than hypocrisy. What parrhesia attempts to do—indeed, *does* when successful—is strip away power's facade of certainty about what it is doing or has done. Truth-telling is therefore analogous to a "solvent" that serves to break down the certainty of power's routinized narratives, self-images, and emotional stabilities. By doing so, it contradicts the rhythmic, psychological, and imaginative strata of aesthetic power, and thus parrhesia creates a high likelihood for the interlocutor to *act* to rectify the dissolving facade of certainty.

The importance of capabilities in a powerful agent's sense of "Self" (its knowledge of its own ontology), which I discussed in the opening chapters of this book, cannot be overlooked. Uncertainty within any "agents" creates a need to act, but uncertainty's presence within the Self of a powerful, especially vitalist actor makes forceful re-actions likely. In this context, what parrhesia engenders is to strip away—even slightly altering—the certainty of this ontology of power, by stimulating action.

Routinization, Rules, and "Tact"

The rules of language and speaking can themselves serve to conceal truth in world politics. I begin here with the work of Nicholas Onuf (1989), which has inspired constructivists to engage how "language is a rule-governed activity" (Wilmer 2003: 221). Rules help construct patterns and structures of language exchanges, and "without these rules, language becomes *meaningless*" (Gould 2003: 61). From the work of Onuf, we recognize that rules do more than set appropriate boundaries for language, as the

> paradigm of political society is aptly named because it links irrevocably the sine qua non of society—the availability, no, the unavoidability of rules—and of politics—the persistence of asymmetric social relations, known otherwise as the condition of rule. (1989: 22)

Rules lead to rule—what Onuf (1989) titles the "rule-rules coupling." Thus, linguistic rules demarcate relations of power and serve to perpetuate the asymmetry of social relations. The *structure* of language games is valued because it provides order and continuity. But because those rules are obeyed so frequently and effortlessly, they are hard to recognize as forms of authority.

Where does the need for such continuity arise? As mentioned in previous chapters, Giddensian sociology suggests that the drive for ontological security, for the securing of self-identity through time, can only be satisfied by the screening out of chaotic everyday events through routines, which are a "central element of the autonomy of the developing individual" (Giddens 1991: 40). Without routines, individuals face chaos, and what Giddens calls the "protective cocoon" of basic trust evaporates (ibid.). Yet, as I have discussed in my other work (2005, 2008a) and as Jennifer Mitzen notes (2006: 364), rigid routines can constrain agents in their ability to learn new information. This is what the rhythmic strata of aesthetic power satisfies. In the context it creates for parrhesia, these routines, connected to an agent's sense of Self, shield that agent from the truth.[4] "The shallowness of our routinized daily existence," Weber once stated, "consists indeed in the fact that the persons who are caught up in it do not become aware, and above all do not wish to become aware, of this partly psychologically, part pragmatically conditioned motley of irreconcilably antagonistic values" (1974: 18). The need for such rhythmic continuity spans all social organizations, including scholarly communities (thus we refer to such communities as "disciplines").

The function of these rules creates a similar problematic faced by the parrhesiastes who is attempting to "shock" these structured rules and habits of the targeted agent. Because the parrhesiastes may find the linguistic rules or at least "styles" or language used by the targeted power to be part of the problem (the notion that one must be "tactful," for instance), she or he must perform a balancing act between two goals. First, the parrhesiastes must challenge the conventions that serve to simplify and even conceal the truth the parrhesiastes is speaking. Second, the parrhesiastes must observe some of these speaking rules, part of which may themselves be responsible for or derivate toward the style of the Self that needs to be challenged by the parrhesiastes. Favoring the first, the parrhesiastes is prone to being ignored as irrational, as someone "on the fringe" or even unintelligible or, in the words of Harry Gould already noted, "meaningless." Favoring the second moves the parrhesiastes away from the truth attempting to be told or at least obscures the truth with the language of nicety. As developed by Epicurean philosopher Philodemus, parrhesia existed within this spectrum: at times, it bordered on "harsh frankness" that was "not mixed with praise"; at other times, the frankness was more subdued (Glad 1996: 41).[5] As the examples of Cynic

---

4. In international politics, during times of conflict, such concealment becomes particularly robust, as Martin Luther King Jr. once posited during the Vietnam conflict: "The time has come for America to hear the truth about this tragic war. *In international conflicts, the truth is hard to come by because most nations are deceived about themselves.* Rationalizations and the incessant search for scapegoats are the psychological cataracts that blind us to our sins. But the day has passed for superficial patriotism. He who lives with untruth lives in spiritual slavery" (http://www.lib.berkeley.edu/MRC/pacificaviet/riversidetranscript.html, emphasis added).

5. Philodemus even proposed a typology of vulnerability to frank speech based on "professions, gender and age"—with such vulnerability being produced for different reasons. For instance, while women are, according to him, more vulnerable because they have "greater psychological insecurity," "famous people and old men, by contrast, are resentful because they think that they are criticized from impure motives and believe that they are wiser than others" (Glad 1996: 34). I find this proposal interesting, the faulty view of gender notwithstanding, because the issue of generational conflict vis-à-vis truth-telling appears again. Note once more, in this vein, Roskin's statement (1974: 567), quoted in chapter 1, that two generations in foreign policy "talk past each other" and that the "experienced" generation sees itself as wiser and therefore less likely to accept criticism from a younger generation.

and academic-intellectual parrhesia provided later in this chapter illustrate, different manifestations of truth-telling as a form of counterpower occupy different spaces along this spectrum—balancing between abiding by these conventions of decorum and style; the need to provide forceful, decloaked truth; or, in the case of Cynic parrhesia, flauntingly contradicting the conventions altogether.

The parrhesiastes will most likely face charges of the first order (ignoring convention) regardless of the manner in which parrhesia is delivered. If, indeed, "the truth hurts" and if the target of such truth cannot deny the facts being delivered, the most convenient option for the victim is to blame "the way" in which the parrhesiastes said something, knowing full well that it was the *substance* of what that person said that was, for the victim, inappropriate or, more to the point, inconvenient.

Tact is one language rule, a collection of embedded or implicit agreements between codiscussants (Giddens 1984: 75–76). Its utility is to smooth out discussions, to provide some structure of sensibility so that substantive issues can be addressed free of "emotional" engagement, although, as tact is based on trust, emotional connections are always just beneath the surface. Jenny Davidson's (2004) fascinating investigation of hypocrisy illuminates the formidable challenge that parrhesia faces in *politeness*. Approaching hypocrisy from a more philosophical level (than the organized hypocrisy literature in IR previously discussed), Davidson demonstrates how habits, traditions, and manners—ideals promoted by conservative writers such as Burke, Hume, and Swift—are diametrically opposed to the truth: "The identification of virtue with politeness renders the ideal of sincerity increasingly problematic, with the effect of polarizing truth and civility" (2). This tact (polite talk), explains Davidson, has "become a form of power that people are willing to acknowledge, giving it a new legitimacy" (168). The "romance with politeness" that was established in the eighteenth century, moreover, is gendered, favoring masculinized notions of discourse.[6] But the problem is more pervasive. Davidson points out that manners are not "exclusively associated with social and cultural conservatism" (173). The so-called multicultural push in the 1990s for "political correctness," while attempting to protect marginalized groups, served to prescribe its own linguistically enshrined boundaries of propriety. Whether facing conventions constructed by the Right, the Center, or the Left, those attempting to use parrhesia confront norms legitimized by a society that rests on a "tyranny of the majority."

---

6. I thank Mlada Bukovansky for leading me to this insight.

Recall the analysis from chapter 1, where I discussed, via Dewey and other aesthetic IR scholars, the importance of aesthetic subjectivity in a community setting. When this is compromised, it is the democratic community that is "wronged," but this same community also recognizes its ability to refashion the Self of power. The parrhesia that has engaged this aesthetic subjectivity insecuritizes more than just the "idea" of this Self. In understanding parrhesia, then, and, moreover, in the obstacle that parrhesia faces, one is able to recognize the precise "danger" a parrhesiastes is placed within (for the general danger is societal disapproval, marginalization, and reaction). Thus, the most basic constraint the would-be parrhesiastes faces is being ostracized by a democratic majority, a semi-"tyrannical" force.

> When once its mind is made up on any question, there are, so to say, no obstacles which can retard, much less halt, its [the Majority's] progress and give it time to hear the wails of those it crushes as it passes. (Tocqueville 1969: 248)

The tyranny of this majority is also satisfied by lying, and as the work of Patricia Owens has identified, "it is tempting and even easy to lie in politics because the lie itself is a form of action. Almost by nature the liar is an actor because the liar wants to change the world from what is to what they want it to be" (2007: 268–69). If this is the case, then the politician, fueled by the need for the majority, is faced with what might be termed an antiparrhesiastic task.

The list of difficulties that a parrhesiastes faces in politics has not been exhausted in this section, of course. But the issues so far explored highlight how difficult truth-telling is in social life and that even the rules that help harmonize social relations serve to mitigate truth. Style and substance matter, and a parrhesiastes who seeks to effectively provoke an interlocutor must take both into account. We have an example of such an account when analyzing a modern-day form of Cynic parrhesia.

III. Cynic Parrhesia

In ancient Greece, the Cynics promoted a view of living life in harmony with nature. Their ethos was to engage the sovereign in a public forum, to state the truth, and to do so free of tact. The enemy of truth, for the Cynic, was found not only in what could or could not be said but in the way it was stated and in the overall conventions of the time (public "decency," forms of dress, etc.). If we return to the informal "spectrum of politeness" discussed in section II, the Cynics were at the most extreme and harsh end. While the parrhesia that existed among cordial acquaintances seeks to "administer just criticism in a temperate way . . . the Cynics went in for stinging reproaches" (Konstan 1996: 12).

Cynic parrhesia included several important components. First, according to Foucault, the spatial aspect of Cynic parrhesia made it a "public activity" or "demonstration" (2001: 108). Truth-telling, as an activity, gathered societal meaning only if it was done out in the open. The Cynics, Foucault tells us, "disliked . . . elitist exclusion and preferred to address a large crowd," speaking in theaters, at feasts, and at athletic events (119). Second, Cynic parrhesia was a particular style, one that concluded "that in order to proclaim the truths . . . that would be accessible to everyone [again, public manner], they thought their teaching had to consist in a very public, visible, *spectacular, provocative,* and sometimes *scandalous* way of life" (117, emphasis added). Thus, the conventions that Cynics flaunted extended past the rules of language—their mere presence and presentation was meant to provoke emotion. Third, the *timing* of parrhesia was important. Foucault discusses how, during this time, parrhesia centralized "the concept of *kairos*—*the* decisive or crucial moment or opportunity" (2001: 111; see also ibid., n. 76). Although he does not elaborate on this point, the literature on Greek *kairos* is voluminous, and, in fact, the timing of speech, this opportune moment, was central in rhetorical understandings from Greek philosophy (Sipiora and Baumlin 2002).[7]

But for what did the parrhesiastes seek timing? This depends on the context of the parrhesiastes' target. In the "temperate" forms of parrhesia where frankness occurs within a regime of friendship, the timing of such speech requires the parrhesiastes to be aware that the truth will hurt a friend and to thus shape the speech accordingly. Plutarch, who was concerned with the differences between a "true friend" and the mendacious flatterer (Foucault 2005: 374), noted how telling the truth to a friend was a "fine art, inasmuch as it is the greatest and most potent medicine in friendship, always needing, however, all care to hit the right occasion" (quoted in Fredrickson 1996: 169). In Cynic parrhesia, however,

---

7. Morgenthau noted that the moment for truth to "maintain itself against power" occurs when society's "opposition to and alienation from the *status quo* becomes so widespread as to undermine confidence in its [power's] cause and the effective cohesion of the 'establishment' itself" (1970: 8).

the search of timing is for the most spectacular moment that will maximize shock, the moment where power is suspended in anticipation.

Cynic parrhesia bases its force on self-sufficiency, where "what you need to have or what you decide to do is dependent on nothing other than yourself" (Foucault 2001: 120). This is a primary element of Cynic parrhesia that qualifies it as a subform of counterpower. The parrhesiastes, in this case, assumes that what he or she says will have an effect because the target of the speech possesses reformative capabilities, but those capabilities vary in content across targets; thus, a parrhesiastes needs to "adapt words to the circumstances of the hearer" (Fredrickson 1996: 169). I use the term *reformative* in the most literal sense, as "reforming"—taking another form from the previous one—could be both positive and negative. Using self-sufficiency as a basis to evaluate the target is a reasonable assumption for the Cynic to make. While Foucault does not clarify why this is so, we might surmise that it is because of the nature of the target itself—the Cynic is using parrhesia against a sovereign or leader, and the target is the ruler. Cynics also used two further techniques: "scandalous behavior"—attitudes that "called into question collective habits, opinions, standards of decency, institutional rules, and so on" (Foucault 2001: 120)—and what Foucault titles the "provocative dialogue."

Armed with these techniques, I think it is possible to reconstruct and make intelligible, in terms of counterpower, the verbal communications issued by al-Qaeda, as a transnational terrorist organization, in playing the role of the Cynic vis-à-vis U.S. hegemony during the past ten years.[8] This is not to suggest that al-Qaeda's provocative dialogue issued in its communiqués are a form of truth-telling. It is, rather, to place the analytical focus on how the kernels of truth that come from the "scoundrels" of twenty-first-century geopolitics can have the effects that they have had (if one shares my conclusion that those effects are real). Broadly, one can consider al-Qaeda similar to ancient Cynics inasmuch as they behave outside the lines of international society and thus represent a group distinct from other organizations, including "terrorist" organizations, such as Hizbollah or Hamas, that seek to become part of international society (Mendelsohn 2005; Löwenheim and Steele 2010).

Osama bin Laden's 29 October 2004 speech is one of many that bin Laden and al-Qaeda have issued over the past dozen years, but it stands out among these others. To begin, it helpfully illustrates the light and

8. See also the argument in Löwenheim and Steele 2010.

rapid condition of counterpower, utilizing the instantaneous functions of postmodern technology. Perhaps of all al-Qaeda communiqués, it most closely resembles Cynic parrhesia. In terms of the spatial and spectacular aspects previously mentioned, the speech was indeed public. Originally posted on an Islamist message forum, it was picked up by the Arab broadcast network Al-Jazeera and almost immediately relayed into broadcasts on the U.S. networks. In terms of spectacle, the image of bin Laden was somewhat striking, as it was the first time he had been seen on video since 2001. Most important, however, was the timing (*kairos*, "moment") of this broadcast, appearing just four days before the 2004 U.S. presidential election.

There is plenty of evidence in the communiqué that bin Laden represented U.S. power to be intensely weak and thought of U.S. hegemony as susceptible to manipulation and quick to re-act. He began by noting not just that U.S. power could be manipulated but that

it was *easy* for us to provoke this administration and to drag it [after us] . . . That was enough to cause generals to rush off to this place, thereby causing America human and financial and political losses, without it accomplishing anything worthy of mention . . . in conducting a war of attrition in our fight with the iniquitous *great power,* [as when] we conducted a war of attrition against Russia with jihad fighters for 10 years until they went bankrupt . . . as a result, they were forced to withdraw in defeat . . . We are continuing the same policy— to make America bleed profusely to the point of bankruptcy, Allah willing. And that is not too difficult for Allah.[9]

This was not just a war of attrition but, specifically, a tactic to use U.S. tendencies, habits, and practices against the West. "The White House policy," bin Laden continued, "appeared to some analysts and diplomats as though we and the White House play as one team to score a goal against the United States of America, even though our intentions differ." Bin Laden also famously stated, "Your security is not in the hands of Kerry or Bush or Al-Qaeda. Your security is in your own hands." Bin Laden reinforced this idea by concluding,

We never imagined that the Commander in Chief of the American armed forces would abandon 50,000 of his citizens in the twin towers

9. "Transcript of bin Laden Speech," http://english.aljazeera.net/NR/exeres/79C6AF22-98FB-4A1C-B21F-2BC36E87F61F.htm (accessed 18 February 2008).

to face this great horror alone when they needed him most. It seemed to him that a girl's story about her goat and its butting was more important than dealing with planes and their "butting" into skyscrapers. This allowed us three times the amount of time needed for the operations, Allah be praised.[10]

What does this all have to do with parrhesia? First, some of bin Laden's charges, as uncomfortable as they might be for some to believe, are, at the core, factually correct. George W. Bush did indeed wait inside a classroom in Florida for seven minutes after being told by White House chief of staff Andrew Card, "America is under attack." The charge by bin Laden does more, however, than simple reporting; it is a taunt, in which, as in Cynic parrhesia, "pride is the main target" (Foucault 2001: 126). The leadership of hegemonic America comes into focus, stripped of its rationally efficient facade and revealed as incompetent, feckless, and even helpless. Bin Laden's mere visual presence, three years after the 9/11 attacks, is enough to enrage. The "flash" of the U.S. Self this manifests is unpleasant. Second, as Foucault mentions, whereas Socratic parrhesia is meant to bring the audience to an understanding or "awareness of ignorance," Cynic parrhesia "is much more like a fight, a battle, or a war, with peaks of great aggressivity and moments of peaceful calm" (2001: 130). This is an incredibly antagonistic speech, but one issued by bin Laden in a style of extreme calm. Perhaps nothing antagonizes an interlocutor more than a brutal message delivered in such a serene manner.

But what should we conclude when we move from bin Laden's speech as simulating Cynic parrhesia to whether the words themselves actually influenced the interlocutor (the American or, more specifically, the American voter)? Here we might temper our conclusions, but there is ample evidence that the tape itself caused a vitalist reaction, one that attests to the Deweyean insights of a strong community-wide emotional reaction to the rupture or break in an aesthetic vision. Overall perceptions at the time indicated that the tape's release and its message would help Bush in his bid for reelection. Bin Laden's tape therefore relates to Cynic parrhesia in another manner—it does not necessarily "bring the interlocutor [the United States] to a new truth," but "it is to lead the interlocutor to *internalize* this parrhesiastic struggle—to fight within himself against his own faults" (Foucault 2001: 133). Commentator Andrew

---

10. Ibid.

Sullivan posited the following on his Web blog hours after the tape's release:

> But why release a tape just before the elections? The obvious impact will be to help Bush. Any reminder of the 9/11 attacks will provoke a national rallying to the commander-in-chief. The *deep emotional bond* so many of us formed with the president back then is Bush's strongest weapon in this election, and OBL has just revived it. The real October Surprise turned out not to be OBL's capture . . . but OBL's resilience. I have a feeling that this will tip the election decisively toward the incumbent. A few hours ago, I thought Kerry was headed for victory. Now I think the opposite. I also have a sinking feeling that that was entirely bin Laden's objective.[11]

John Kerry himself reflected on this issue the following summer on the U.S. talk program *Meet the Press*, stating, "We were rising in the polls up until the last day when the tape appeared. We flat-lined the day the tape appeared and went down on Monday."[12] Finally, as recorded in Ron Suskind's recent account (2006: 40), the view on the day of the tape's release at CIA headquarters was that it would boost Bush, as evidenced by Director John MacLaughlin's statement that "Bin Laden certainly did a nice favor today for the President."

We do not, alas, have clear polling data from which to substantiate Sullivan's reaction, so I do not feel completely comfortable concluding that bin Laden's message was solely responsible for Bush's reelection victory, let alone that bin Laden intended to manipulate a skittish American public into reelecting Bush for another four years. The most conclusive analysis one can use in this context comes from a recent study conducted by Cohen and others (2005). The study was an experiment conducted to assess whether a key assumption of terror management theory—namely, "mortality salience," or a vision that reminded people of their own mortality—increased support for George W. Bush. The study concluded that "Bush was favored over Kerry following a reminder of death" (178).

---

11. Available at http://time-blog.com/daily_dish/index php?dish_inc=archives/ 2004_10_01_dish_archive.html#10990896141116968. One anonymous Bush campaign official was reported to state that the bin Laden tape was "a little gift" to the Bush campaign (http://www.prisonplanet.com/articles/november2004/011104tapeboostsbush.htm).

12. "Kerry Says bin Laden Tape Gave Bush a Lift," *New York Times*, 31 January 2005, http://www.nytimes.com/2005/01/31/politics/31kerry.html.

Yet the bin Laden tape, in a counterpower sense, was about a bit more than just mortality—it was also about aesthetic integrity. Sullivan is perhaps closest to the mark in the preceding quote—discussing how bin Laden challenged the "emotional bond" (psychological stratum) Americans had formed with Bush (most likely as a convenient proxy for the idea of "America"). The tape was a manipulation of U.S. strength, and it is an illustration of the assertion made in chapter 1 that the psychological conditions of trauma make such moments of counterpower possible (but not predictable). Notice how neoconservative commentator David Brooks, almost on cue, wrote in his 30 October *New York Times* column,

> Here was this deranged killer spreading absurd theories about the American monarchy and threatening to murder more of us unless we do what he says. One felt all the old emotions. Who does he think he is, and *who does he think we are?* One of the crucial issues of this election is, Which candidate fundamentally gets the evil represented by this man? Which of these two guys understands it deep in his gut—not just in his brain or in his policy statements, but who *feels it so deep in his soul that it consumes him?* It's quite clear from the polls that most Americans fundamentally think Bush does get this.[13]

In the conclusion of his Berkeley lecture on Cynic parrhesia, Foucault relates the story of one Cynic, Diogenes, engaging the king Alexander the Great as reported in the Fourth Discourse of Dio Chrysostum.[14] Alexander participates in the encounter because Diogenes, as a Cynic, is a challenge to his authority. According to Foucault,

> Diogenes wants to hurt Alexander's *pride* . . . [and] at the beginning of the exchange, Diogenes calls Alexander a bastard and tells him that someone who claims to be a king is not so very different from a child who, after winning a game, puts a crown on his head and declares that he is king. Of course, all that is not very pleasant for Alexander to hear. But that's Diogenes' game: hitting his interlocutor's pride, forcing him to recognize that he is not what he claims to be. (2001: 126, emphasis added)

---

13. "The Osama Litmus Test," *New York Times*, 30 October 2004, http://www.nytimes.com/2004/10/30/opinion/30brooks.html?_r=1&oref=login&hp&oref=slogin).

14. The account of the engagement is "a fictional dialogue," but it is still an illustrative way to end this section, in my view.

Foucault suggests that the encounter represents "a struggle occurring between two kinds of power: political power [Alexander] and the power of truth [Diogenes]." Its purpose was not to provide the Cynic with any kingly possessions but, rather, to force the interlocutor "to fight a spiritual war within himself" (ibid., 133). If David Brooks was correct, then the bin Laden message forced Americans to ask, "Who does he think *we* are?"—invoking perhaps the most important of those "old emotions" all Americans confronted. In this way, the U.S. War on Terror is no longer about the what of al-Qaeda but is now about the who, the Self, the aesthetic ontology of America. Such an internal struggle remains unresolved to this day.

IV. Academic-Intellectual Parrhesia

The remainder of this chapter investigates one further site where we can assess the prospects for parrhesia: in academia. I say this appearance is "possible," if not probable, because of the constraints academics and all intellectuals have faced in recognizing their roles as parrhesiastes, the degree to which their work can appear as parrhesiastic, and the aesthetic nature of parsimonious theory—the type of theory that most easily melds with power (see Steele 2007a; Oren 2006). As the Kuhnian discussion in chapter 1 suggested, academics face a multitude of conventions. Sometimes these conventions are necessary—we must couch our arguments and results in the vessels of a language that resonates with our audience. There is no "pure" form of intellectual parrhesia. But the conventions faced by the academic-intellectual that I am most interested in exploring in this section are those that are connected to power—specifically the pervasive power of a hegemonic actor (nation-state) and the role of power in constructing the basis for knowledge, against and within which the academic is situated. "The important thing here," Foucault writes, is that "truth isn't outside power, or lacking in power" but "a thing of this world . . . Each society has its own regime of power" (1980: 131). The academic works within this regime.

I qualify my remarks here in a couple respects. First, it is not my position that academics should be focused on producing and codifying "the truth." We should be and often are aware of the myriad ways in which social reality itself is produced through images and discourse. Our scholarship should speak to uncovering these processes so that they are no longer taken for granted as objective social facts (Cox 1986). Academics should imbue their work with a permanently skeptical purpose, focusing on how the "truths" of politics are produced and conversely challenged, as well as (as opposed to solely) focusing on how those social facts produce observable patterns of behavior.

Second, I understand how academics may wish to make the world a "better place," and so it is understandable (if not problematic) when they align their work with power. But I also aver that certain theoretical dispositions will recognize better than others the manner in which power contaminates the truth. Indeed, this was the point I made when discussing the differences between liberal idealist and constructivist IR perspectives (Steele 2007a), the former being more blind to the "spillover" of subjects and objects in the social sciences. More immediately, I also assert that academic-intellectuals can practice a form of parrhesia by becoming more self-aware of how and when our scholarship coincides and aligns with power relations—when it, even minimally, seems to abed those in power with a sense of legitimacy, scientific validity, or authority. A central goal of such a parrhesiastic investigation is to identify how and when such validity becomes authoritative.[15] My hypothesis is that the closer academics get to power, the more they are tantalized by its ability to change outcomes, and thus the more impaired they become in their ability to identify the streams of social reality they examine in their work. The truth of such depictions of social reality become contaminated. Once contaminated, of course, such academics are probably too far removed from the truth to speak it and too co-opted to extricate themselves from that position. So it will be left to others to speak this truth not only against the power they examine but also against the academics who help reify it through their "objective" conventions, rules, and habits.

Constraints on Academic-Intellectual Parrhesia

To begin, what most budding scholars of all disciplines (IR included) face, of course, is the fear of unemployment, on the one hand, and the

---

15. One philosophical choice, for Foucault (although the one he did not choose), was to "opt for a critical philosophy which will present itself as an analytic philosophy of truth in general." This coincides with the purpose noted in the previous paragraph. The other choice, the one that "founded the form of reflection" Foucault worked within, was to "opt for a form of critical thought which will be an ontology of ourselves, an ontology of the actuality," or an "ontology of the present" ("What Is Revolution?" in Foucault 2007: 95). This is the goal expressed in this paragraph.

need to be unconventional, on the other. This is what James Rosenau (2004) once asserted, regarding his role as a mentor to budding IR scholars, as the dialectic between "courage" and "caution."

> It is an excruciating question because, on the one hand, encouraging one's students to conform to the established orthodoxy is anathema for me, but, on the other hand, one is reluctant to have them risk their careers because they listened to you and got set back because they broke with the prevailing orthodoxy prematurely. (515)

The "danger" that the academic faces exists even before they begin speaking "truth to power." For both Foucault (2001: 15) and Morgenthau,[16] the way that we know a parrhesiastes is sincere is that she or he is willing to put herself or himself in danger. Orthodox research is produced for the powerful, it is picked up, protected, reproduced, and reinforced by these regimes of power. What partially frees an academic from the concern expressed by Rosenau is, of course, tenure. In this sense, as Morgenthau notes, the U.S. government could have made his life quite a challenge: "The White House could threaten me with the FBI and make the Internal Revenue Service waste many man-hours in repeated audits of my tax return . . . But it could not deprive me of my livelihood or of my freedom to speak and write" (1970: 16).

Of course, even with tenure (or, before that, in order to get tenure), academics must accept a myriad of conventions, "rules" of analysis, in order to get their work accepted (published, praised, assigned, etc.). One challenge even tenured academics face is the disapproval of the society in which—sometimes toward and against which—they work. This has become particularly perceptible when one examines the role of the academic-intellectual during times of war. While there are many examples that can be used to illustrate the challenge U.S. academics face in practicing parrhesia, I focus here on three contexts: World War I; the Vietnam War (specifically addressed by Hans Morgenthau); and, in brief discussion, the current U.S.-led War on Terror.

Dewey's work on aesthetics has been presented in chapter 1. He is also instructive for his role as an academic-intellectual in the first half of the twentieth century. Dewey broke with many of his fellow academics by supporting U.S. participation in World War I. Dewey's support for the

---

16. As Morgenthau wrote, "that power should be limited by truth is an intolerable affront" to those in power (1970: 21).

war stemmed from his belief that a "better" social order could be actuated through this participation in the war in Europe. Indeed, American participation was assumed to be vital toward making that war a transformative one where international democracy was realized. In a "time of national hesitation," Dewey (1917) summarized his view on World War I as follows:

> I have been a thorough and complete sympathizer with the part played by this country in this war and I have wished . . . for its successful prosecution. As has been said over and over again, this is not merely a war of armies, this is a war of peoples.

The position of those who supported U.S. participation in World War I in general and Dewey's position and experience in particular is noteworthy for four reasons. First, it demonstrates how intellectuals can be seduced by the possibility of realizing their own ideals through war, a process that has been repeated time and again through the relationship between U.S. academics and their government. Second, Dewey's alliance with power demonstrates how idealism originating from the Left can resort to nefarious actions that resemble the more recent constricting of academic freedom the American Right has engaged in since 9/11.[17] Dewey most notably attempted to silence Randolph Bourne, his chief critic, by getting him removed from the editorial board of the periodical *Dial*, which was Bourne's primary medium of critique (Westbrook 1991: 212).

Third, Dewey's support of the war and penning of that support through his work shows how academics can be caught "off guard" when their ideals do not get properly promoted, executed, or effused by power. In this regard, it is important to note how quickly Dewey's "growing second thoughts about his own commitment to the war" materialized, second thoughts fueled by a realization after Versailles that "Americans had put themselves in a position of having fought a war that did little to advance the cause [Dewey's cause] of international democracy" (Westbrook 1991: 239–40). This is especially noteworthy because Dewey is but one of a long line of liberals who have noted their dismay with power's ignorant, if not inefficient, executions in carrying out their liberal ideals, culminating with modern-day liberal idealists' angst regarding George W. Bush's use of the democratic peace to justify the foreign policy adventure in Iraq (Russett 2005).

---

17. For an overview of such clamps, see Wilson 2005.

A fourth issue is illustrated by Randolph Bourne's critique of Dewey. When academics fuse their "truths" with the exercise of power, those truths themselves become contaminated.[18] They are no longer truths to be told but, instead, become authoritative and used to control others. The original truth is used by the unseduced to invigorate their critiques, which makes possible academic-intellectual parrhesia. Bourne could critique Dewey by turning the latter's "philosophy back upon" Dewey "with a vengeance," hoisting Dewey "with his own petard," pointing out where Dewey fell "prey to the very mistakes his philosophy was designed to prevent" (Westbrook 1991: 197, 206, 203). Chief of these mistakes was the "ideological blindness" that Dewey pursued once the war began.

Hans Morgenthau faced a similar position during the Vietnam years, as did all academics who, like him, opposed U.S. involvement. Morgenthau had, in his words, "undertaken the naïve assumption that if power were only made to see the truth, it would follow that lead," but when he brought his views on the war to U.S. president Lyndon Johnson, the latter engaged thereafter in a "systematic attempt, making full use of the informal powers of his office, to discredit and silence the voice of the dissenter [Morgenthau]" (1970: 4). Johnson furthermore had the assistance of the Freedom House organization. Freedom House produced a statement, published in the 30 November 1966 *New York Times,* which, in Morgenthau's words, branded those like him who opposed the Vietnam War "extremist and irresponsible critics." More interestingly, Morgenthau and his fellow war opponents were disciplined by the Freedom House letter for the *manner* in which they expressed their criticisms. According to Morgenthau, the document "tells us [academics opposing the Vietnam War] that we are morally entitled to criticize the government, but not with regard to the fundamental issues it enumerates. That is to say, we are not morally entitled to criticize the government in any meaningful way" ([1967] 1970: 49). IR scholars might note that this is the same organization that soon thereafter (in 1972) began rating countries' levels of political and civil rights—ratings that have been used by numerous IR analysts in their voluminous work on, among other creations, the democratic peace. Within the academy, as reported by Richard Ned Lebow (2003: 240), this opposition came at the expense of the presidency of the American Political Science Association, whose "conservative pro-war administrator quietly mobilized pro-war professors

18. Bourne claimed that Dewey had been "sucked into the councils at Washington and war organizations everywhere" (quoted in Kaplan 1956: 260).

to block his nomination." In this case, Morgenthau faced the dangerous position of a parrhesiastes, or as Muriel Cozette so eloquently puts it, "Morgenthau then certainly spoke truth to power, and paid the price for doing so" (2008: 17).[19]

Moving to the near-present day, which academic-intellectuals might we analyze as participating in power's "regimes of truth" since 9/11? I will try to avoid "naming names" in this section, but I do identify here a spectrum of participation on which those seduced by power can be placed. On one end, we can identify psychologists who were used in "a way for C.I.A. officials to skirt measures such as the Convention Against Torture" in their interrogations of al-Qaeda suspect Abu Zubaydah, which imbued a sense of social "scientific legitimacy" to the torture techniques.[20] Close to this position, we place the "intellectuals of statecraft" who have generated what Francois Debrix titles "tabloid discourses" of war (2007: esp. chap. 3).[21] Also on this end, we can consign the "experts" on Arab culture consulted by the Bush administration prior to the Iraq invasion, individuals who told the latter, "The Arabs cannot pull themselves out of their historic rut. They need to be jolted out by some foreign-born shock. The overthrow of the Iraqi regime would provide one" (Packer 2005: 51).

Further along this spectrum of seduction, we can place authors such as Jean Bethke Elshtain, who have carefully but assertively imbued the Iraq War as one where, "on balance," most "criteria of just war were met" (2004: 183). Rather than outright support for the means of war, Elshtain uses the criteria here to provide for the justification of the U.S. War on Terror's stated "ends," although what those ends might be is never clear, as I discuss in the following chapter. Alison Howell's (2007) study on the impact the "psy" disciplines have had on the subjectivity of detainees held by the United States as deranged or irrational illuminates why these perspectives also participate in serving the interests of power. Moving

---

19. For the impact of Vietnam on the U.S. academy and the development of conventional security studies, see Walt 1991: 216.

20. See Katherine Eban, "Rorschach and Awe," *Vanity Fair,* July 2007, http://www.vanityfair.com/politics/features/2007/07/torture200707; Jane Mayer, "The Black Sites," *New Yorker,* 13 August 2007, http://www.newyorker.com/reporting/2007/08/13/070813fa_fact_mayer.

21. Here as well, we can include the legal experts who provided counsel to the Bush administration regarding its coercion regime—individuals such as John Yoo and Jay S. Bybee. I save a more detailed discussion of their role in this regime until the next chapter, where I discuss their participation in providing a legal justification, through constitutional reinterpretation, for such coercion.

even further along the spectrum, we can (perhaps contentiously) place the democratic peace scholars who, by providing an "empirical law" cloaked in correlative "truth," enabled the Bush administration to deploy a narrative of democratization used to stem criticism of its otherwise disastrous Iraq policy.[22]

In this vein, a fundamental example of academic-intellectual parrhesia has recently been proffered by Tony Smith (2007). Smith, a self-described "liberal internationalist," sheds light on various points of the previously mentioned spectrum. He notes several groups, including democratic peace scholars, comparative analysts studying democratization in the Middle East (134–37), and neoliberal jurists (169–72), all of whom made the case that democracy was a universal whose spread should be the foundation for U.S. foreign policy. Smith compellingly points to what he titles the "neoliberal hawks" who "gave this signature feature [democratic peace] of the Bush doctrine the *gravitas* it came to possess" (164).

Smith's book bears on parrhesiastic force in several ways. First, like the parrhesiastes Foucault (2001: 82) mentions, Smith here demonstrates the "courage to oppose the demos" or group—in this case, his fellow liberal internationalists. Those named in Smith's book have expressed dismay that Smith would attempt, in his other work, to "purge [his] fellow liberals."[23] Interestingly enough, this liberal dismay is obviously, at its heart, concerned with the underlying message, rather than its style. Yet the publicly expressed "disappointment" the targeted neoliberals have with work like Smith's is that it is "too partisan." We thus see how parrhesia faces power's disapproval regarding its style, its lack of tact. This diversionary technique must be recognized for what it is—an attempt by defensive individuals to avoid the elephant in the room, rather than a noble claim for mature decorum.

Second, academic-intellectual parrhesia like Smith's book helps to, as Morgenthau once remarked, "lay bare what is wrong," which "is not an exercise in ex post facto fault-finding. Rather it is an act of public

---

22. The quote "empirical law" is from Levy (1988: 662). See Steele 2007a for the general critique of democratic peace "science"; see also Ish-Shalom 2006 and Kratochwil 2006. Prominent democratic peace scholar R. J. Rummel is perhaps the most assertive defender of policies to promote democratization, concluding his final post on his blog with the statement that "the most important thing we can do for humanity is this: *Promote democratic freedom everywhere.*" Also interestingly, Rummel has argued that the tenure system that Morgenthau noted (as mentioned earlier in text) should be abolished in favor of five-year contracts.

23. Slaughter 2007.

purification and rectification" ([1968] 1970: 416). Even if the parrhesiastes is "sacrificed" in danger, his or her words perform this function of rectitude and holding the participants in regimes of power to account. If such a function is marginalized—if it is delegitimized because of "concerns" over tone rather than substance, for instance—then "faults, undiscovered and uncorrected, are bound to call forth new disasters" (ibid.). Or if we are to remain using neutral language, such regimes of power will continue to reinforce their basis for domination.

Why is this important? Is it that big of a deal when academics stand with the power they are positioned to challenge? I recognize the skepticism regarding the idea that intellectuals can play any role in abetting the regimes of power, and I am at times quite sympathetic to the opinion that what academic-intellectuals say and write is really only important to the small clans in which they work. Surely those on one side of the spectrum, such as the psychologists who explicitly enabled the torture of Zubaydah, are most responsible here. But what is at issue is to conjecture how and why any of these individuals along this spectrum become allied with the causes of power.

How do academic-intellectuals become allied with power? The terrain on this issue is fortunately well traveled. One manner is through the creation of "experts" (Buger and Gadinger 2007), where the specialization of knowledge makes possible the metanarratives that construct conditions for authority. What Foucault titles the "specific" intellectual is an "expert" who "has at his disposal, whether in the service of the State or against it, powers which can either benefit or irrevocably destroy life" (1980: 129). Morgenthau posited that this expertization was an artifact of liberal philosophy, which sought to replace politics with the efficiency of expert administrations (1946: 29). In his later work, Morgenthau noted how an

> intellectual can enter the political sphere as an *expert*. As such, he does not question the purposes and processes of government . . . [Instead] he tells the powers-that-be what they need to know in order to achieve a particular result and how they must go about achieving it. Given a certain objective, of which he may or may not approve, he gives advice as to how to achieve it . . . He puts his truth at the service of power. (1970: 16, emphasis added)

This expertization of international politics leads to rating and ranking. Oded Löwenheim's impressive work on international "examination periods" avers how Freedom House and Transparency International have performed a governmentality function, rating and ranking so that the effect of such judgments is a normalizing one. These organizations'

> indices and surveys not only reflect perceptions and opinions of allegedly objective experts, but are also highly imbued with normative assessments and ideological judgments which are subtly embedded in the methodology and questionnaires used to produced them. (2007b: 10)

Again, I acknowledge the caution skeptical readers might express with the notion that academics play much of a role, if any at all, in such regimes of governmentality. Yet what else might we call these regimes other than a partial representation of Walter Lippman's (later recanted) vision of a "Good Society" where the statesman "would be little more than an 'intermediary between the experts and his constituency'" (Kaplan 1956: 358)? Löwenheim's work, like Smith's, is itself a form of academic-intellectual parrhesia, as it adeptly excavates the manner in which academics serve to participate within, indeed reinforce, a global regime of governmental authority, in order to "disturb people's mental habits," as Foucault once stated (1988: 265).

We can identify at least three reasons why academic-intellectuals align their arguments with power—why they would mesh their "will to truth" with such power structures. First, many academic-intellectuals are idealists, and they court power (as power courts them) in order to realize these ideals. We might posit that there is no such thing as "realism," that competing idealisms surround us, and that the only persons who are not idealists are those who are in power. Such a broad definition of idealism can be supplemented with a more narrow conception I take here—that idealism is the belief in humankind's ability to progress.[24] It is, more to the point, an attempt to "assist [the] march of progress to overcome ignorance," as Hedley Bull (1972: 34) described the "students of international relations" after World War I; or as Morgenthau once remarked, it is the belief that "international peace and order [can] be achieved

---

24. For the distinction between idealism in IR theory and idealism in social theory, see Wendt's detailed discussion (1999: 24–25) . In short, Wendt posits that idealism rests on the "minimal claim" that "the deep structure of society is constituted by ideas rather than material forces" (25). Vibeke Schou Tjalve (2008) has made a recent compelling case that Niebuhr and Morgenthau were, at root, progressives as well—albeit a case tempered with a profound skepticism and embracing of contingency.

through scientific precision and predictability in understanding and manipulating international affairs" ([1964] 1970: 251). The purpose of progressive academic-intellectuals is to identify the contaminating conditions of ignorance that get in the way of progress and then to seek their removal from the social sphere they study. This was most clearly at work regarding Dewey's posture toward World War I, when, in the words of Sidney Kaplan, power's use needed to be "intellectualized": "by April 1917, [Dewey's] creative intelligence . . . was fully committed to American entrance into the war" (1956). Here academic-intellectuals are attracted to power's ability to eliminate the conditions of contamination they have reified in their research as the most problematic ones in social life.

Somewhat related, but even more pressing, is the second reason for why academic-intellectuals align themselves with power: it seduces. Morgenthau (1970: 24; see also Cozette 2008: 17) identified three factors that account for this susceptibility of the intellectual community. The first he titles "the conformism of American Society," and while he asserts that this conformism is peculiar to the United States because it lacks the "historic, ethnic and cultural values" that form the basis of other countries' societies and thus has to "rely for its survival upon unorganized social pressures to keep its members in line," his additional position is that "to be a 'regular fellow' or a 'member of the team'" is *not* a virtue unique to Americans alone (Morgenthau 1970: 24). The need for societal cohesion, the maintenance of order, forms the basis of all horizontal disciplines.

The third factor for how intellectual communities remain silent and become corrupted is "personal ambition": "Academics in particular, that is, intellectuals, who are professionally committed to the pursuit of truth, are not immune from aspirations for power, academic and political" (Morgenthau 1970: 25). We might assume here that academics realize this ambition by speaking *for* power. Yet Morgenthau argues that academics have balanced their commitments to truth and power by dealing "with 'safe' subjects in a 'safe' manner," that is, by remaining silent on "the great issues of political life." Thus, we notice here how the appeal to the rules of positivism intends to avoid the "sources of bias and inefficiency created by methodologically unreflective research designs" (King, Keohane, and Verba 1994: 229). Such appeal, moreover, allows us to gather data, or "use the facts we know to learn about facts we do not know" (ibid., 46). Yet as Kratochwil has so forcefully pointed out, the idea of value-neutral "facts" is itself problematic.[25] As Murielle Cozette writes, the reason why the academic is "courted" by power is because power must "hide under the guise of morality in order to be effective, and for its action to appear legitimate" (2007: 21). Power thus embeds values within scientifically "objective" studies. Thus, the social scientist imbues the values of a particular community with a sense of neutrality.

For this reason, a fourth factor regarding the seduction of power illuminates how academic silence is abetted and reinforced by the government's

> varied, subtle, and insidious instruments with which to forge reliable ties with large segments of the intellectual world. Thus the "military-industrial complex" . . . is duplicated by an academic-political complex in which the interests of the government are inextricably intertwined with the interests of large groups of academics. (Morgenthau 1970: 25)

If we are to accept that truth-telling is a function necessary for the academic-intellectual, we should heed Morgenthau's argument that such individuals must continuously perform a critical dislodging function, perpetually opposed to power. Cozette explains, "Political science is almost inevitably at odds with the demands of society, as it unmasks power, and necessarily goes against conventions which are socially contingent" (2008: 18).

This is easier said than done, for there is also a largely underappreciated aesthetic embedded in these conventions and the theories that become co-opted by power. When we take notice that the value of "parsimony" is often used coterminously with the word *elegant*[26] and that parsimonius models are those that tend to be most useful for the political justification of power (see Steele 2007: 47), this means that an academic-intellectual parrhesiastes is engaging not only power but an aesthetic connection between a community of academics and a particular theory that may defy rational explication. Such parrhesia will expose a

---

25. "It is not simple observable elements (natural kinds) but *values* that determine the operations of our concepts . . . What counts as an instance of, for example, war or democracy, is neither answerable by a closer look at the phenomena, nor is it solved by strict operationalization and inter-coder reliability" (Kratochwil 2006: 9–10).

26. Although he is not making a case for parsimony, Schweller has used the term *elegant* in describing Waltz's theory in at least two studies (2003: 312; 1998: 10).

"hole" in at least the normative underpinnings of a theory, but it will also create a likely emotional backlash. Yet this should not surprise us, for as previous sections of this book have attempted to demonstrate, power (including that of academic groups who become allied with it) is also vulnerable and thus can be engaged and confronted by those who are brave enough to do so.

V. The Glimmer of Parrhesia

When facing these dire warnings regarding the manner in which academic-intellectuals are seduced by power, what prospects exist for parrhesia? How can academic-intellectuals speak "truth to power"? It should be noted, first, that the academic-intellectual's primary purpose should not be to re-create a program to replace power or even to develop a "research program that could be employed by students of world politics," as Robert Keohane (1989: 173) once advised the legions of the International Studies Association. Because academics are denied the "full truth" from the powerful, Foucault states,

> we must avoid a trap into which governments would want intellectuals to fall (and often they do): "Put yourself in our place and tell us what you would do." This is not a question in which one has to answer. To make a decision on any matter requires a knowledge of the facts refused us, an analysis of the situation we aren't allowed to make. There's the trap. (2001: 453)[27]

This means that any alternative order we might provide, this hypothetical "research program of our own," will also become imbued with authority and used for mechanisms of control, a matter I return to in the concluding chapter of this book.

When linked to a theme of counterpower, academic-intellectual parrhesia suggests, instead, that the academic should use his or her pulpit, their position in society, to be a "friend" "who plays the role of a parrhe-

---

27. Morgenthau has a different view on this issue, asserting that the academic's words "make very little direct difference in the real world": "Thus I could criticize the Kennedy Administration for its handling of the Cuban missile crisis because I had no decision to make . . . President Kennedy, on another occasion, commented upon my criticism of one of his policies by saying that I ought to sit where he did. President Kennedy had a point" (1970: 17). More than this, Morgenthau also disagrees that the general public is "denied" information the government already has; in fact, what "impedes sound judgment," he says, is the overall quantity of information that the government has access to (19).

siastes, of a truth-teller" (2001: 134).[28] When speaking of then-president Lyndon Johnson, Morgenthau gave a bit more dramatic and less amiable take that contained the same sense of urgency.

> What the President needs, then, is an intellectual father-confessor, who dares to remind him of the brittleness of power, of its arrogance and blindness, of its limits and pitfalls; who tells him how empires rise, decline and fall, how power turns to folly, empires to ashes. He ought to listen to that voice and tremble. (1970: 28)

The primary purpose of the academic-intellectual is therefore not to just effect a moment of counterpower through parrhesia, let alone stimulate that heroic process whereby power realizes the error of its ways. So those who are skeptical that academics ever really, regarding the social sciences, make "that big of a difference" are missing the point. As we bear witness to what unfolds in front of us and collectively analyze the testimony of that which happened before us, the purpose of the academic is to "tell the story" of what actually happens, to document and faithfully capture both history's events and context. "The intellectuals of America," Morgenthau wrote, "can do only one thing: live by the standard of truth that is their peculiar responsibility as intellectuals and by which men of power will *ultimately* be judged as well" (1970: 28). This will take time,[29] but if this happens, if we seek to uncover and practice telling the truth free from the "tact," "rules," and seduction that constrain its telling, then, as Arendt notes, "humanly speaking, no more is required, and no more can reasonably be asked, for this planet to remain a place fit for human habitation" ([1964] 2006: 233).

Earlier sections of this book described the contours of what is meant by "counterpower" in international relations—those micropressures that

---

28. The role of a "friend" or ally using parrhesia seems to be the basis for Turkish filmmaker Bahadir Ozdener's comments defending his controversial film *Valley of the Wolves of Iraq*, which depicts several scenes of U.S. forces committing atrocities in modern-day Iraq. Ozdener stated, "Turkey and America are allies, but Turkey wants to say something to its friend. We want to say the bitter truth" (http://news.bbc.co.uk/2/hi/entertainment/4700154.stm).

29. "Nothing is more inconsistent than a political regime that is indifferent to the truth; but nothing is more dangerous than a political system that claims to prescribe the truth. The function of 'free speech' doesn't have to take legal form . . . The task of speaking the truth is an infinite labor: to respect it in its complexity is an obligation that no power can afford to shortchange, unless it would impose the silence of slavery" (Foucault 2001: 464).

power inevitably faces because of its vulnerability. In the current case, U.S. power has been understood as an aesthetic ontology that anchors individuals to their nation, and it is this routinized stream of images that disciplines American citizens and allows the politically powerful to continue the status quo. Yet what "America" is can be challenged, reformed, and manipulated both in the model of the Cynics (al-Qaeda) or in the model of the academic-intellectual.

On an individual level, a truth-teller who exercises counterpower could be a friend who informs us over lunch of an aesthetic observation—we have food on our face, we have gained some weight, or we look fabulous. Internationally, the parrhesiastes is attempting to do the same thing (for good or ill)—to get an interlocutor to engage its own sense of Self, to *sense* its aesthetic Self, and, by doing so, to transform not just that Self but the eyes that gaze upon it, which see it in a new light. For this parrhesiastes, there is a danger that we should not easily overlook. It could be as miniscule as a moment of uncomfortable silence, a more serious imperiling of a friendship, or, finally, being ostracized by an academic or political community. But there are benefits to truth-telling as well. As the following chapter intuits, parrhesia is a counterpower that can also make possible other forms. Truth-telling itself can bring the Self into a new light, or it can serve to excavate the protective visual layers of a targeted power and open up the process to which I now turn—what I call "self-interrogative imaging," where the beauty of power's Self is put on trial through the display of de-aestheticized images.

FOUR

The Image of Power:
Self-Interrogative Imaging

Four daguerreotype likenesses of my head were taken, two of them jointly with the head of Mr. Bacon. All hideous . . .

At seven this morning Mr. Bacon came and I went with him to the Shadow Shop, where three more Daguerreotype likeness were taken of me, no better than those of yesterday. They are all too true to the original.

—FORMER PRESIDENT JOHN QUINCY ADAMS, UPON SEEING
HIS PHOTOGRAPHED IMAGES, IN TWO DIARY ENTRIES
DATED 1 AND 2 AUGUST 1843 (KRAINIK 2002: 24)

This chapter advances a third form of counterpower: *self-interrogative imaging,* or the process by which the aesthetic basis of power is reflected, refracted, and reoriented through images. Images imprint and objectify the "glory" of group Selves and are representations of those Selves. Self-interrogative imaging refers to the distribution of images that represent (and re-present) the subjectivity of power's aesthetic basis. Both complimentary and critical forms of self-interrogative imaging spring forth to manipulate power by, like other forms of counterpower, engaging its aesthetically constructed sense of (in this case) national "Self." Such images are not necessarily concerted "strategies"—they manifest spontaneously, like other forms of counterpower.

This chapter presents four case illustrations of such self-interrogative imaging. Two sets come from the Vietnam War: the series of photographs taken by Ronald L. Haeberle documenting the 1968 My Lai massacre and the photo of the April 1975 Fall of Saigon. The other two case illustrations come from the U.S.-led Iraq War: the images of the burned contractors hanging from a Fallujah bridge in late March and early April 2004 and the series of Abu Ghraib photographs made public in late April of the same year. Several aspects of these counterpower illustrations make them especially important for study. Most pressingly, these all represent iconic moments in U.S. foreign policy history, although they by no means exhaust the illustrations that could have been chosen for these cases. The My Lai and Saigon examples, as I discuss in more detail later in this chapter, are, in certain ways, microcosms of the "tragedies" that a collective generation of Americans have "viewed" politics through for four decades. Abu Ghraib and Fallujah are important moments of the U.S. war in Iraq. They were, in many ways, turning points in the U.S. effort in terms of losing the "moral" (Abu Ghraib) and "strategic" (Fallujah) high grounds at a similar point in the war (the spring of 2004). While I investigate the impact of the images, I am also interested in the discursive interventions of power—how they attempted to create meanings of these images and to what degree those meanings successfully disciplined each of these four instances of counterpower. As one will notice, however, the images were varied in their counterpower positioning. In certain cases, such as the My Lai photos, the images were quickly archived and kept hidden from U.S. memory. Representing deaestheticized moments in the Self of U.S. power, they needed to be isolated, concealed, and forgotten. In other cases, such as the Fall of Saigon photo, such images made little initial impact but were then fit into ideological narratives such as those of the neoconservative movement for the purpose of engaging the issues of "strength and will" into a vitalist American identity, once its large generational cohort acquired, beginning in the 1980s and thereafter, the resources to develop such a narrative.

In all cases, self-interrogative imaging is perhaps the most vibrantly generationally shaped form of counterpower. Because images are *experiential*, they are like codes of data imprinted on the collective minds of generations. This does not mean, of course, that each generation interprets these images in the same fashion. Indeed, what makes these images interrogative is that they are sites of political contestation of the Self of U.S. power. They disturb precisely because they challenge an aesthetic vision of that national Self, a vision that is always politically negotiated.

Another part of the process of self-interrogative imaging comes about because of the ambiguity of power's operational environment. The important trope of "darkness" has pervaded analyses of two of the cases that follow (My Lai and Abu Ghraib). I am interested in just what or who is darkened here. To reset my argument from earlier chapters, the basis for power's influence lies in its ability to keep hidden, to obfuscate, to make ambiguous, and to conceal. Here again we intersect with Foucault's late work on aesthetics, which sees the power of pleasure, love, and eros in the *concealing* of the body. What makes the attraction to the "painting" of the idea that is America so powerful for citizens is this ambiguity. It allows the citizen to "love" his or her country in a deeply personal manner. Foucault speaks of how powerful eros is when it occurs "in the dark." The object of affection "could be wounded by the unloveliness of the images . . . The cruel image can be an excellent means of protecting oneself against passion or even a means of ridding oneself of it" (Foucault 1984: 138). Therefore, if power is "most effective when masked," as Hans Morgenthau believed (Lebow 2003: 232),[1] then self-interrogative imaging's dilution of a targeted power comes from its ability to unmask.

But as will become more evident in the case illustrations that follow, this ambiguity also *makes possible the actions* that, when revealed, rapidly shock the ontology of power. It makes possible, in other words, the forms of resistance or counterpower that "spring from" the "constitution of an organized multiplicity" of power (Foucault 1977: 219). This darkness is the perfect laboratory for psychological connections to power, but in such a laboratory, power's malfeasance runs wild.

Two Tragedies: Vietnam and the Hole in U.S. Power

It is hard to overstate the importance of Vietnam to the so-called Baby Boom Generation of Americans. Michael Roskin, a scholar whose work was engaged in earlier chapters, writing in this conflict's wake (1974), averred that Vietnam was producing an "emerging noninterventionist orientation . . . [that] urge[s] *limitation* of American activity" (575, emphasis added). With the benefit of thirty-five years of hindsight, we can conclude that regarding Vietnam's impact, Roskin had it half-right. Vietnam is collectively engaged by the Baby Boom Generation, but its point of view is bitterly splintered. It is true that certain baby boomers accept

---

1. In another vein, there is perhaps something to be said for the "excitability" of power in the first place, and it reveals an intersection between post-structural and classical realist approaches to IR. Foucault discusses the approach of the Stoics in reducing (but not eliminating) the "pleasure inscribed by Nature": "the task therefore is to elide pleasure as a sought-after object" (1984: 139). As I have mentioned elsewhere, a key article of what I have termed "reflexive realism" is "prudence-as-stoicism," an article that was always present in classical realist philosophy (2007b: 279). I return to this argument in the conclusion.

the lesson of limitation, or nonintervention, from Vietnam. But Roskin collects quotes from at-the-time liberals such as John Kenneth Galbraith and Arthur Schlesinger Jr., and he was also writing this article before the Fall of Saigon (575–76). Thus, he had yet to observe the emotionally charged and vitalist reaction from the emerging neoconservative movement.

A recent journalistic account of the experience of 2008 U.S. presidential candidate John McCain proves instructive. McCain's Vietnam experience is well known. A Navy pilot whose plane was shot down in 1967, he was captured and taken as a prisoner of war soon thereafter. He spent six years in a Hanoi prison, was tortured by his Vietnamese captors, and returned to the United States in 1973. McCain's intense experience was therefore different from that of some of his fellow senators who are Vietnam veterans (such as John Kerry of Massachusetts, Chuck Hagel of Nebraska, and Jim Webb of Virginia), who "found themselves unable to recognize their enemy in the confusion of the jungle," whereas McCain's contact with his torturers was much more "black and white" (Bai 2008).[2] The point is that McCain and many baby boomers, while all collectively engaging Vietnam as a counteraesthetic experience, as a "hole" in the idea that is the U.S. Self, have nevertheless taken different lessons from that experience in terms of what type of challenge it provided to the U.S. Self.[3]

Thus, Vietnam represents "two tragedies." In one, it was the tragedy of forcing U.S. soldiers into a jungle of hazy enemy targets and missions, for an amorphous purpose only loosely connected to the national interest. In the other, "the lesson McCain and other conservatives took away from this version of history is that America was driven from Vietnam principally because the voters, discouraged by dire reports from a skeptical media, lost their *will*" (Bai 2008, emphasis added). Two events documented with images captured and were refracted through the U.S. Self in Vietnam: the massacre at the hamlet of My Lai and the 1975 U.S. withdrawal from Saigon. These are both, I suggest, political sites where the basis for future U.S. actions were reflected on and interrogated. They

---

2. Vietnam veteran and former Georgia senator Max Cleland stated, in this same article, that while McCain was his "friend and brother," he also thought that "you learn something fighting on the ground, like me and John Kerry and Chuck Hagel did in Vietnam" (Bai 2008).

3. Bacevich calls this collective group the "forces of reaction," in terms of revising military institutions in the wake of Vietnam (2005: 6).

were, in a related sense, the apparitions of de-aestheticized Selves that the U.S. pushed away from in future decades.

My Lai

The notes of a 21 November 1969 telephone conversation between national security advisor Henry Kissinger and secretary of defense Melvyn Laird detail an emerging political problem for the Nixon administration.[4] Kissinger called Laird to talk "about the atrocity case; the President wants to make sure L [Laird] got on top of that—got a game plan." Tellingly, Laird asked Kissinger if he had seen the pictures of the massacre. Kissinger said, "No . . . should I?" to which Laird responded that Kissinger "might as well not. They're pretty terrible." Laird said, "He's like [*sic*] to sweep the whole thing under the rug," but the pictures were already, by this time, "out there." The two men recognized the inherent lack of control that they and, ostensibly, U.S. decision makers had over the photos. The pictures belonged to a member of the platoon that committed many of the atrocities, and by the time of their conversation, they had already been published in the *Cleveland Plain Dealer*. What is interesting is that the pictures took some twenty months to fully publicly emerge. One scholar points out,

> What occurred during the course of November 1969 was what the US military *had feared for eight months*, ever since Ronald Ridenhour's account of the killings at My Lai had arrived at the Pentagon, the loosening of its *control* not just over the distribution of knowledge about the atrocity, but also over the process of investigation and prosecution, and over its own reputation and the reputation of the war it was fighting in Vietnam. (Oliver 2006: 49, emphasis added)

The background and events of the My Lai massacre have been well documented, but a brief synopsis is appropriate here. On the morning of 16 March 1968, members of Charlie Company from the Twenty-third Infantry Division entered the village of Son My, which consisted of four hamlets (My Lai, Tu Cung, My Khe, and Co Luy). Encountering no enemy fire, the soldiers killed mainly women, children, and elderly men.

---

4. Kissinger and Secretary of Defense Melvin Laird, 21 November 1969, 3:50 p.m., Nixon Presidential Materials Project, Henry A. Kissinger Telephone Conversations Transcripts, Chronological File, box 3, file 3, 083–084, Washington, DC.

One officer in particular, Lieutenant William Calley, is thought to have rounded up and killed nearly eighty civilians by himself (Oliver 2006). Many of the dead were deposited into ditches. In one harrowing scene, soldiers who were eating their lunch in the aftermath walked over to one of the ditches to "finish off" survivors who were still moving and thus distracting the soldiers from finishing their meals (Anderson 1998: 37–38). All told, between 374 to 504 Vietnamese civilians were killed in My Lai and the surrounding areas.[5] A soldier in Charlie Company, Ronald Haeberle, took several photographs of the massacre as it unfolded, some with an Army camera, which were confiscated, and others with his own camera, which would be developed in color.

One officer, Hugh C. Thompson, flew his helicopter over the area during the last stages of the massacre and saw the large number of civilian bodies lying throughout the village. Some thirty years after the massacre, Thompson remarked,

> We came across a ditch that had a lot of bodies in it and a lot of movement in it. I landed and asked a sergeant there if he could help them out . . . He said the only way he could help them was to help them out of their misery, I believe. I was shocked . . . I thought he was joking . . . As we took off and broke away from it, my crew chief, I guess it was, said, "My God, he's firing into the ditch." We had asked for help twice, actually three times by then. Every time the people [My Lai civilians] had been killed. We were not helping out these people by asking for help. (Anderson 1998: 29)

Thompson then landed his helicopter and personally lifted surviving villagers out of the ditches and into safe areas outside of My Lai. In one rescue instance, he ordered his door gunner, Specialist Lawrence Colburn, to target his gun on American soldiers nearby while they helped more villagers into the helicopter. When the day was over, Thompson reported what he saw to his commanding officer, Major Frederic Watke, who then relayed the information up the chain of command. However, while this led to an investigation conducted with what Oliver terms "a politic [of] discretion and economy of energy" (Oliver 2006: 92), the inquiry ended shortly after the massacre. It would be left to others, such as Seymour Hersh and Ronald Ridenhour, to engage the events in a way that forced

---

5. The former number is the figure the U.S. military calculated after its internal investigation. The latter number is cited by Vietnamese officials.

the Pentagon and eventually the U.S. media to reopen an investigation into exactly what had happened that day in My Lai.

The following passage comes from the letter Ridenhour wrote on 29 March 1969 and sent to Arizona congressman Morris Udall, the Pentagon, and the State Department.

> Exactly what did, in fact, occur in the village of "Pinkville" in March, 1968 I do not know for *certain*, but I am convinced that it was something very black indeed. I remain irrevocably persuaded that if you and I do truly believe in the principles . . . that form the very backbone that this country is founded on, then we must press forward a widespread and public investigation of this matter with all our combined efforts . . . I hope that you will launch an investigation immediately and keep me informed of your progress. If you cannot, then I don't know what other course of action to take.[6]

In part because of this letter (and, of course, the work of Seymour Hersh),[7] the military began an internal investigation into the events, at that time almost a year old. Several elements of Ridenhour's story and biography indicate that this is a good example of the parrhesia analyzed in the previous chapter. Ridenhour was an infantryman serving in an area close to My Lai, and while he was not putting himself in any physical danger by this letter, he was compromising the bond of "honor" that exists among members of the military, and he risked being ostracized by the U.S. "demos" as a whole. Before writing the letter, but after he had interviewed eyewitnesses to the massacre, Ridenhour relates that he "went home and talked to my friends and my relatives and all of the people who I thought had been my mentors. They all, almost to a person, said 'Shut up. Shut up. This is none of your business. Leave it alone'" (Anderson 1998: 40). Some thirty years after writing the letter, Ridenhour reflected that he "was a kid. I had no idea how to do it, but I knew the first thing I needed was the facts" (38).

Indeed, Ridenhour's letter relays eyewitness accounts of the massacre

---

6. Anderson 1998: 205.

7. Hersh 1970, 1972. Ridenhour and Hersh did "connect," and Ridenhour indeed helped fuel Hersh's investigation of the massacre as well (Anderson 1998: 40). William George Eckhart, while noting that Ridenhour wrote an "articulate letter," asserts that "it was only after the Army Inspector General had completed his investigation . . . that journalist Seymour Hersh reported the incident" (2002: 314). Eckhart fails to note, however, that Ridenhour's letter helped loosen the Pentagon into investigating the incident in the first place.

in an almost methodical and detached manner, at least until the closing paragraph (previously quoted). As mentioned in the opening chapters to this book, forms of counterpower can work on one another, making other forms of counterpower possible. In this situation, Ridenhour's par-rhesia helped solventize power,[8] opening an investigation where pho-tographs of the massacre were uncovered, photographs that proved vital to upending the aesthetic integrity of the U.S. Self.

Seymour Hersh's investigation was published as a story in thirty-five newspapers on 13 November 1969. But even then, the reaction was not marked, as by "mid-November 1969 there were three stories with more urgent claims upon the front pages of the national press": the second American moon landing; an antiwar rally in Washington, D.C.; and Vice President Spiro Agnew's remark that news commentators represented an "elite liberal bias" (Oliver 2006: 44). The print version of My Lai was represented as simply something that "happened" all the time in a war like Vietnam. Tellingly, in light of what I discussed in the last chapter, NBC anchor David Brinkley remarked at this time that "the full truth is not yet obtainable" (ibid., 46).

Then Ronald Haeberle took his photographs to the *Cleveland Plain Dealer*. The regional paper decided to publish them on 20 November 1969. How important were these images? Oliver asserts that "the dam finally broke, as narratives of the massacre and images of its victims flooded the national media" (2006: 47). The *Dealer* itself posted the im-age of twenty dead My Lai civilians lying on a road on its front page—an image that included naked civilians and even children. That evening, the *CBS Evening News* "exhibited the photographs, one after another, *in complete silence,*" and the pictures were purchased by *Life* magazine for the right to display them in color (ibid., 48, emphasis added).

Significant about this particular moment (20 November 2008) is how the photographs, assisted by Hersh's more detailed second account (which included eyewitness reports), were presented in an unvarnished fashion. They were displayed in the *Dealer* and in silence on the *CBS Evening News* without any immediate comment from U.S. policymakers. The images were there for all to see on this Friday in the United States, where the gaze of the U.S. citizen could be its own judge, and it is in this moment that the aesthetic of U.S. power was suspended and fragile. Noteworthy as well was a study published a few years after the story broke

---

8. Tellingly, an award titled the "Ridenhour Prize for Truth-Telling" was founded in 2003. See http://www.ridenhour.org/.

(Thompson, Clarke, and Dinitz 1974) that indicated that participants who saw the images "were more condemnatory of the My Lai-killings" than those who were presented with simply a written description.[9] As Oliver also notes, even though they were not evidence of Lieutenant William Calley's part in the massacre, the Army prosecutor presented the pictures, "enlarged and in color," during Calley's trial.[10]

It is therefore apparent that the images had much *independent* weight as they interrogated the Self of U.S. power. In this way, they represent the broader insecurity in narcissistic power in general and in the U.S. experiences with Vietnam specifically. The impact of the photos is, somewhat problematically, not about the Vietnamese Other but, instead, a form of "self-absorption . . . in which the tragedy of the war resonates primarily in its consequences [including aesthetic consequences] for Americans" (Oliver 2006: 4). Yet the images were distributed in a certain American context. Here again, *kairos,* or "the moment" of counterpower, was important. Timing was crucial here, in that what exacerbated the impact of the images was a more raw American journalistic culture, where "the discourse of 'taste' which had previously limited pictoral representations" had been loosened by 1969 (51).

As the correspondence between Laird and Kissinger reported at the beginning of this section indicates, the pictures had a strong initial impact on the American public, especially among the media and commentariat, indicating the extent to which My Lai at least initially represented a rupture in U.S. self-perception. An editorial in the *New York Times* on 22 November 1968 remarked that My Lai "may turn out to have been one of this nation's most ignoble hours" (quoted in Oliver 2006: 2). Following Lieutenant Calley's trial, Reinhold Niebuhr remarked, "This is a moment *of truth* when we realize that we are not a virtuous nation" (quoted in McFadden 1971, emphasis added).

The My Lai pictures interrogate the Self of power in both an immediate and a deeper sense. In the immediate sense, much like the Abu Ghraib images I discuss later in this chapter, the pictures reveal a very undisciplined manifestation of U.S. power. Elderly men, women, and children lay dead as the result not of "collateral damage" from the air

---

9. The study mentioned, furthermore, that responses became even more extreme the more often the images were seen by participants.

10. The pictures of My Lai can be seen at various Web sites. During the fortieth anniversary of the massacre, newspapers around the United States republished the pictures. See, for instance, http://www.salem-news.com/articles/march162008/my_lai_remembered_3-15-08.php (accessed June 2009).

but of bullets shot at point-blank range. The images are tragic, of course, but they are also immediately ugly—naked and bloody. The images of My Lai implicate the subjectivity of U.S. power more deeply as well. They "articulated the need for American society to confront the true reality of itself, as a prelude either to reform or to the fashioning of a new, more complex national self-knowledge" (Oliver 2006: 131).

Yet what has to be refashioned? Must the Self look into these images and gain a broader sense of empathy for the (in this case, Vietnamese) Other? No. Instead, what has to be refashioned is the ontology of the Self. Will it be, as Oliver indicates, the U.S. Self that appeared "in the coverage of the first two moon landings" (2006: 131)? Or is it the vision of My Lai? While an aesthetic smoothness for the Self never can truly exist, a sense of extreme insecurity generated by My Lai forces the Self to re-act. One should note as well the origin of the pictures. They were *produced* (as the Abu Ghraib photos were) not by a crime scene investigator who was attempting to bring justice to the victims by gathering facts but by someone who was participating or at least a bystander to the crime itself. The U.S. looks at these crimes in these images through "the gaze of the killers" (Oliver 2006: 132–33). To return to the metaphor that Foucault gives us about the painter viewing a painting and the latter working back along the former, these images are thus an object (rather than subjects) on which the interrogation of the Self occurs. They are the easel and canvas where Self has been painted, an ugly scene that requires attention.

Additionally, as I have already mentioned, the production of power to the point where it is repulsed against its aesthetic Self results from a lacuna, a lack, of definitional provisions regarding its own operation. It operates "in the dark"; while the aesthetic Self of power is imagined, we only see glimpses of its contours when it breaches what I term, in the next chapter, the "transgressional limit." This indicts both the tactical and strategic goals being pursued in Vietnam and operates along several streams of ambiguity. As it was connected to the amorphous "domino theory" of the spread of communism, the goal of soldiers in Vietnam was to "win." To use such a sophomoric concept as an operational goal makes sense in a sporting event, where points and a time clock are kept to ensure finality. But what did "winning" mean in Vietnam?[11] The words of William Calley are illustrative here.

---

11. The concept of winning is not just a vague operational goal for the military. As Peter Feaver ([2005] 2006) and others have famously posited, the most important determinant of democratic public opinion regarding conflict is not casualties but the perception of "winning" or "losing."

I think of Mylai and say, *My god. Whatever inspired me to do it?* But truthfully: there was no other way. America's motto there in Vietnam is "Win in Vietnam," and in Mylai there was no other way to do it. America had to *kill everyone there.* (Anderson 1998: 117)

One can immediately see how such an interpretation of "winning" would produce outcomes that would implicate the aesthetic integrity of the U.S. Self. But, indeed, this operational goal led to the tactical objective of using "body counts" as a mark of progress. Stephen Ambrose appropriately commented that "using the body count as a way to measure progress in the war was a terrible idea, and it had all kinds of awful repercussions . . . When the situation is that you cannot dispose of an enemy until you have killed him, you have the opposite of progress" (Anderson 1998: 117). Thus, ambiguous wartime goals contain both a power and an inherent fragility.

As we will notice later regarding the streams of ambiguity that made Abu Ghraib possible, the concept of "winning" in a conflict is inherently problematic in a post–World War II era. But whereas "body counts" were used in Vietnam as a metric of progress, distinctively nebulous and even more problematic goals—gathering of information and the "demonstration" of authority to an Arab Other—were formulated to "win" the hazily titled War on Terror.

Strength and Will: Saigon, 1975

Precisely five years before the last helicopters left Saigon, President Richard Nixon explicated what was "at stake" for U.S. interests in Southeast Asia, justifying the U.S. incursion into Cambodia in April 1970 with these words:

If, when the chips are down, the world's most powerful nation, the United States of America, acts like a pitiful, helpless giant, the forces of totalitarianism and anarchy will threaten free nations and free institutions throughout the world. It is not our power but our *will and character* that is being tested tonight. The question all Americans must ask and answer tonight is this: Does the richest and strongest nation in the history of the world have the *character* to meet a direct challenge by a group which rejects every effort to win a just peace . . . ? If we fail to meet this challenge, all other nations will be on notice that despite its overwhelming power the United States, when a real crisis comes, will be found wanting. (Semple 1970)

The Paris Peace Accords of January 1973 provided for a cease-fire between the North and South Vietnamese and made way for the withdrawal of U.S. combat forces in Vietnam. By the time hostilities between the North and South had largely resumed later that year, with the Communist forces capturing more territory along the way, most U.S. combat forces had left.

The weeks leading up to the Fall of Saigon illustrate the chaos that surrounded U.S. policymakers and the trauma felt in the United States over losing, in the words of a *Newsweek* column from that time, "its first foreign war" (Mathews and Elfin 1975: 18).[12] U.S. president Gerald Ford's behavior was most demonstrative of this chaos, in his attempts throughout that month to reassure the American people that South Vietnam would still maintain its independence. But accounts from that time recognized that these were empty words. In light of what Nixon had claimed in 1970, Ford, in one instance, had "felt compelled to warn the rest of the world five times during his 44-minute press conference that America had not suddenly turned into a pitiful, helpless giant" (ibid.), although Ford had not used those exact words.

By mid-April of 1975, with Communist forces turning toward Saigon, South Vietnamese president Nguyen Van Thieu resigned, but not before blaming the loss of South Vietnam on the lack of U.S. aid, which he claimed was promised to him as a condition of the Paris Peace Accords (Browne 1975). On 16 April, the United States began withdrawing five thousand diplomatic and military support personnel, with the remaining one thousand Americans leaving, largely through airlift to the U.S. ships of the Seventh Fleet, in the last days before Communist forces overtook Saigon.

On the day of Saigon's fall, President Ford stated that the withdrawal "closes a chapter in the American experience . . . I ask all Americans to close ranks, to avoid recrimination about the past."[13] But perhaps more important, Americans awoke that day and saw in their morning papers perhaps the iconic image of the Fall of Saigon. The picture is of the CIA Air America Huey helicopter evacuating the last remaining American personnel out of Vietnam. Most tragically, a long line of Vietnamese civilians, too many to fit into that one Huey helicopter, stand waiting to board—lest they, being connected to American interests in Vietnam, be subjected to the whim of Communist forces streaming into the city

---

12. Interestingly, the article was titled "Spectacle of Defeat" (*Newsweek*, 14 April 1975, 18).

13. Finney 1975: 85.

around them.[14] The *New York Times* described the scene in the picture as "a crewman from an American helicopter helping evacuees to the top of a building in Saigon for flight to a U.S. carrier" (Finney 1975: 86). The *Economist* titled its article covering the withdrawal "Operation Panic."

The reactions to the withdrawal and its images were, of course, not as pronounced as the My Lai images, but the impact was still palpable. A *New York Times* editorial around this time commented,

> The United States left Vietnam with the same confusion and lack of direction that took this country there in the first place. The scenes of agony and tumult in Saigon yesterday, as the helicopters lifted American diplomats and panic-stricken Vietnamese away, add up to one more sorrowful episode at the conclusion of an American—and Vietnamese—tragedy.

But the editorial went on, somewhat presciently, to state that the "meaning" of the withdrawal would still leave "too many questions . . . unanswered in the heat of defeat; too many others will be *deliberately obscured in the days—and years—to come,* for the *protection of reputations and ideals that will not easily be given up*" ("The Americans Depart," *New York Times,* 30 April 1975, 37, emphasis added).

"Never in American history," neoconservative Norman Podhoretz recently wrote, in an article tellingly titled "America the Ugly" (2007), "had our honor been so besmirched as it was by the manner of our [Vietnam] withdrawal."[15] For the neoconservative and even for other analysts who recognize its import, this image is a hole in the beauty of the U.S. Self.[16] Several elements of weakness are at work in the photo—disorder,

---

14. The photo can be viewed at http://www.worldsfamousphotos.com/the-fall-of-saigon-1975.html (accessed June 2009).

15. Podhoretz, "America the Ugly," *Wall Street Journal,* 11 September 2007, http://opinionjournal.com/extra/?id=110010589. Recall from chapter 2 that this is the same Podhoretz who admitted his admiration of the "Negro's . . . superior physical grace and beauty" (1963: 99).

16. Fareed Zakaria, surely no neoconservative, recently asserted, "There could not have been a worse time for America than the end of the Vietnam War, with helicopters lifting people off the roof of the Saigon embassy" ("Beyond Bush," *Newsweek,* 11 June 2007). A *Guardian* columnist remarked, "In 1975 the Americans suffered a spectacular military defeat at the hands of North Vietnam and the Vietcong, with US helicopters seeking to rescue leading US personnel from the tops of buildings as Vietnamese guerrillas closed in on the centre of Saigon. It was to shape American foreign policy—in particular, a desire to avoid overseas military entanglements—for decades. Indeed, the rise of the neoconservatives was partly predicated on a rejection of what they saw as American defeatism during and after the Vietnam war" (Martin Jacques, "The Neocons Have Finished What the Vietcong Started," *Guardian,* 8 December 2006, 39).

flight, and, most important, the inability to help out the Vietnamese allies who had presumably stood by the Americans during their Vietnam
participation. What "winning" would have entailed in Vietnam is still unclear to many historians, but what losing looked like was captured in this
photo. Also of interest was the myth that the building that served as the
location of the evacuation was the U.S. embassy. On the contrary, the
Dutch photographer who took the picture wrote, on the thirty-year anniversary, "The photo is not of the embassy at all; the helicopter was actually on the roof of an apartment building in downtown Saigon where
senior Central Intelligence Agency employees were housed."[17]

It is striking and even impressive how revisionist "historians" have
served to challenge the notion that Vietnam was a lost cause from the
start—to access the image of Saigon as an example of a policy that could
have ended differently. One need not go any further than the titles of
these works—*Stolen Valor, Triumph Forsaken, Lost Victory*, for example—to
assess how much the American perception of Vietnam challenges Americans themselves. Notice that these titles imply that the goal of "winning"
was not only reachable but identifiable. The targets of these critiques are
many and include journalists, such as David Halberstam, Peter Arnet,
and Neil Sheehan, who were stationed in Vietnam and "fabricated" accounts of American and South Vietnamese battlefield losses (see Moyar
2006); the failure of Congress to have the will to continue aiding the
South Vietnamese army after the United States began to draw down
forces after 1971 (esp. William Colby 1989); and the peace movement as
a whole (Sorley 1999).

But while David Campbell accurately remarks that neoconservatives
swept into this breech by appealing to "past virtues and earlier dangers,"
such attempts hardly provided the "order out of chaos" that they were intended to capture (1998: 162). In fact, such heroic narratives of neoconservatism served to destabilize a decadent post-Vietnam American order—to engage this traumatic blow to U.S. self-perception of strength, in
order to move forward into a new era where U.S. ideals could be asserted, to justify this assertion abroad with a critique from within. As I discuss in the following subsection, the timing of the Fall of Saigon's interrogative image can be partially indexed to the rise of this incoming
generation of neoconservatives. Of course, these narratives were compelling because they resonated not just with a few neoconservatives but

---

17. Hubert Van Es, "Thirty Years at 300 Millimeters," *New York Times*, 29 April 2005.

with a U.S. populace and especially a generation, as a response to the hole of Vietnam. I return to this issue in discussing the Iraq case sets.

Reflecting (and Refracting) Strategy

What is of central consequence to note about these two illustrations is how inversely related they are in terms of their trajectorial impact on the American psyche. The substantial shock from My Lai waned rather quickly. The "evidence . . . dried up" (Anderson 1998: 46), and the issue of American atrocities during the Vietnam War were papered over during the "conservative drift of national memory" in the 1980s and 1990s, a drift that focused on the U.S. Self as a victim from Vietnam rather than an "executioner" in Vietnam (Oliver 2006: 4). Initially, the memory of Saigon was also encouraged to be "buried," as the quotes by Ford and Kissinger indicate. But over time, the place of My Lai as an image of the U.S. Self during Vietnam has been challenged by the "lost honor" image of Saigon, 1975, with the latter image being politically resurrected as a visual prop for the national greatness narratives of the neoconservatives.[18]

How could this shift have occurred? From the standpoint of the burying of My Lai, it probably began in the weeks and months that followed the dissemination of the photos. We might term one strategy the "quarantining" or "bad apple-ing" of some of the soldiers in Charlie Company, isolating them as deviant exceptions that were wholly responsible for the atrocities. The approach taken here is a very scientific and specific one: first, find out what causes individuals such as Calley or Captain Ernest Medina (Calley's commanding officer) to have the propensity to commit such atrocities; then, seek to remove those soldiers from duty. As Oliver notes, "the army command and the Nixon administration had a com-

---

18. Admittedly, there was one U.S. institution that did not "forget" My Lai: the military. My Lai (and the Vietnam War in general) forced the U.S. military to reform itself in several ways following the end of U.S. participation in that conflict. This was necessary, as Carey McWilliams notes (1972: 144), because the U.S. military would have an enormous difficulty recruiting if prospective soldiers thought that they would be put in a position like those soldiers who were in the My Lai massacre. In light of this, besides an investigation and punishment against Calley, the military became an all-volunteer force that prided itself on "professionalism" (Ambrose in Anderson 1998: 118). But My Lai itself remained salient in American military culture at least through the 1991 Persian Gulf War, when commanders reportedly stated to their soldiers, "No My Lais in this Division—Do you hear me?" (Eckhardt 2000).

mon investment in establishing a distinction between the actions of Charlie company and the broader ethos of American war-fighting in Vietnam" (2006: 73), for recruitment and morale would suffer even more than they already had during a time (the early 1970s) when the institution of the U.S. military was taking a lot of black eyes.

The isolation of Calley and Medina as a "few bad apples" was also reinforced by the dissemination of information testifying to the heroic humanitarianism of the majority of soldiers in the U.S. military (Self). This was indexed by a demonizing of the Vietnamese guerilla Other, focusing on the atrocities committed by the Viet Cong against the South Vietnamese civilians (Oliver 2006: 76). Oliver's characterization of "the efforts of the Nixon administration to *classify* the massacre as *something apart,* and to keep it *segregated* from discussion of wider national policy in Vietnam" (ibid., emphasis added), echoes the views of Foucault's many works (1965, 1977) on the power of classification and the ability of power (in this case, in the form of the Nixon administration) to isolate deviance in order to control it.

Thus, initially, by trying Calley and finding an objective, identifiable fault for My Lai, power began repainting the aesthetic integrity of the U.S. Self—or at least the integrity of its military forces—in the wake of the My Lai story's breaking. It was much easier to characterize My Lai as an aberration rather than a logical production of the U.S. Self. Of course, as I intuited earlier, what produced My Lai was the ambiguity of the "body count" policy (what counted or did not count as an enemy "kill"), and the broader American reaction to My Lai implies that the public assumed that these types of atrocities were a "part of war," especially a war against an indistinct enemy. Throughout the 1980s, articles in U.S. newspapers would carry the opinions of baby boomer politicians who saw the "essential error on the American side" in Vietnam being the failure "to use decisive force early enough" (Lelyveld 1985: 29). These lessons would become especially instructive for this movement when it came into power in the early part of the recent decade.

The U.S. War on Terror

One should note the striking manner in which neoconservatives have used Vietnam to support the continued U.S. presence in Iraq.[19] Perhaps

19. Non-Western media sources have recognized the import of the Fall of Saigon. As Packer notes (2007), on the thirtieth anniversary of the fall, Al Jazeera reran footage of the evacuation of Saigon.

most ironic was President Bush's use in August 2007 of Graham Greene's *The Quiet American* character Alden Pyle as an example of the "American purpose . . . and dangerous naivete" that led, in Bush's view, to the tragic U.S. withdrawal from Vietnam and even the atrocities of the Khmer Rouge in Cambodia.[20] Who knew that such "naivete" inherent in the withdrawal would lead to genocide in a neighboring country (a genocide, by the by, that was ended by the intervention of Vietnamese forces)? Of course, the misappropriation of Greene's Pyle was shocking—Pyle, as readers of Greene's novel know, was an American whose idealism in 1955 led him to advocate *more* of an American presence in Vietnam, rather than "withdrawal." But Bush's judgment that U.S. withdrawal precipitated the Cambodian genocide and other post-1975 atrocities is an assessment that goes back to at least 1980, when Charles Horner (like John McCain, a member of the air force who flew missions during the Vietnam War) made the same case in an article appearing in *Commentary* magazine (Horner 1980). It was this "defeat," this loss, that fueled a generation to "win" the next major conflicts. While ambiguous (as I already mentioned), the goal of "winning" has, as Fierke notes, been common for many nation-states, especially after a past loss—which in essence forces "a continuing need to repeat the past in order to do it differently in the future" (2007: 197). The problem arises, of course, when the past is redone in a future that is itself, a priori, also different. Thus, the "doing" is a reflection of an internal sense of insecurity, and, again, the Other is an object through which the "mirror stage" of the insecuritized subject of power can be cleansed.

Fallujah and Disorder

By March 2004, the United States, in fighting Operation Iraqi Freedom, was facing a raging insurgency in the so-called Sunni Triangle of Iraq, which included the cities of Ramadi, Samarra, Balad, Baqubah, and Fallujah. It was in the latter where an iconic image would lead to a turning point in the U.S. war effort—an image of two U.S. contractors hung from a bridge after being beaten, burned, and dismembered, with Iraqis celebrating nearby.

Shortly before these events, U.S. marines took over responsibility of Fallujah from the Army's Eighty-second Airborne Division. Along with

---

20. "Remarks at VFW convention," 22 August 2007, http://www.whitehouse.gov/news/releases/2007/08/20070822-3.html.

responsibility, the marines intended to bring a new approach to rooting out the insurgents in the area, attempting to more rigorously engage "both the people and the enemy" (Ricks 2006: 331). This approach was meant to "wean Fallujahns from the insurgency" (Ballard 2006: 11). But on 31 March 2004, security personnel from Blackwater, a contracting firm, entered the city to patrol a route that a contractor convoy would take the following day. They did this by running a marine roadblock. The two sport-utility vehicles they road in were ambushed by insurgents.

The images here were disseminated through an added dimension. While it is not exactly clear which news outlet covered the event, the visions of the Fallujah bridge were televised and carried on major U.S. cable networks such as CNN (Ricks 2007: 332). Importantly, the images "provoked a powerful response down the chain of command, starting from Washington, where the images of Mulsim mobs burning Americans evoked memories of October 1993 in Mogadishu, Somalia" (ibid.). Indeed, a LexisNexis search of print media accounts of the Fallujah bridge uprising produces over one hundred "hits" when the words *Fallujah* and *Somalia* or *Mogadishu* are included. Then State Department spokesman Adam Ereli was asked at a 1 April 2004 press conference whether the incident would track with "the Mogadishu precedent that people will lose, sort of, the support or the stomach for staying in Iraq"? Ereli responded by stating, unequivocally, "The Mogadishu precedent was that, following attacks, we left. And that's not going to be—that's not going to happen here, I can tell you right now."[21]

Intellectuals joined the chorus of interpretations as well. Mark Bowden, author of the Somalia chronicle *Black Hawk Down,* wrote an article for the *Wall Street Journal* detailing, as it was titled, "the lessons of Mogadishu." The vitality of "strength and will" as a deterrent in the face of such "barbarity" compelled the United States to recognize that "the worst answer the U.S. can make to such a message—which is precisely what we did in Mogadishu—is [to] back down."[22] By invoking the "haunting" images of Mogadishu, commentators were more concerned with the ghosts of American ontologies past: weakness in the face of barbarity, lost "will" in the face of a challenge, a "cut-and-run" strategy that

21. "Daily Press Briefing, 1 April 2004," http://www.state.gov/r/pa/prs/dpb/2004/31043.htm.

22. Bowden, "The Lessons of Mogadishu," *Wall Street Journal,* 5 April 2004, http://www.opinionjournal.com/editorial/feature.html?id=110004911.

led to a perception that American democracy lacked the fortitude to "keep fighting."[23]

The pictures were again a proper form of counterpower. Light and rapid, they caught a U.S. public off guard. The U.S. marines in the area pushed for a more orderly and cautious response to the atrocities (Ricks 2007: 332), but as Bing West described it in his account of the images, they were "sickening," and according to those in Washington, "the humiliation was no longer a battlefield crime but a symbol of American humiliation," and thus the response was "emotional and aggressive" (2005: 59).

So U.S. power re-acted against the visions of current and past perceived instances of lost will and, most important, an image of weakness. The U.S. marines were ordered to launch an assault—appropriately (for the purpose of the mission) dubbed "Operation Vigilant Resolve." The operation included over twenty-five hundred marines along with artillery and tanks entering the city in a slow-rolling assault. Yet in a (seemingly) bizarre twist, shortly after it began, an order came on 9 April 2008 for an American cease-fire (Ricks 2007: 342). This led to a siegelike stalemate of the city and an eventual pullback of the U.S. forces. The city was handed over to the so-called Fallujah Brigade, which, according to Ricks, "had far more in common with the insurgents than they ever would with the Marines" and was led by a former general in Saddam Hussein's Iraqi Republican Guard. Ricks notes that this ending "was one of the lowest points of the entire U.S. military effort in Iraq" (2007: 343).

Yet Ricks perhaps misses the point. If the objective is not to root out insurgents or even mainly vengeance but, rather, to demonstrate *resolve* in the moment when the aesthetic Self of subjective power is questioned and to re-create that Self's aesthetic integrity, then the operation was, at least, a qualified success. Once such resolve has been demonstrated, the attack can be called off. The militarily strategic or even tactical utility of such a demonstration is *beside the point*. Of course, like the policy of detainee abuses to extract information—like any ambiguous goal of U.S. aesthetic power—such an emphasis on vitality and strength facilitates even more challenges. But it is hard to consider these consequences when you're being chased by ghosts of a past Self.

---

23. The manner in which the Mogadishu incident has haunted U.S. foreign policy is well documented. See Dauber 2001.

Abu Ghraib and Detainee Abuse

Detainee abuses have been committed in several theaters of the U.S. War on Terror, primarily at the centers in Guatanamo Bay, Cuba; Bagram Air Base in Afghanistan; and, most especially, Abu Ghraib prison in Iraq. Beginning with Operation Enduring Freedom, the Bush administration has maintained that the Geneva Conventions did not apply to individuals captured fighting for the Taliban or al-Qaeda—"the former on the grounds that Afghanistan was at the time a failed state, the latter because al Qaeda is a non-state actor."[24] However, whatever loopholes this allowed interrogators to jump through concerning the treatment of these two (nonexclusive) groups of individuals, the same practices were applied "to prisoners in Iraq" who were neither the products of a failed state nor nonstate agents. Thus, the same techniques that were used in Abu Ghraib in 2004 had first been used in Guatanamo Bay.[25]

Following the 9/11 attacks, the Bush administration's legal counsels in the Justice Department went about redefining torture—requiring that the *intent* to inflict suffering had to be equal to "serious physical injury," such as organ failure or death (Bybee and Yoo [2002] 2005: 153–61). There are two opaque elements at work in such a policy. While the narrowing of torture to suffering at the level of serious physical injury was an unprecedented rerendering of the definition of torture, it was the slide to "intent" that was most problematic to "prove" in a court of law. Systematically and through multiple levels, U.S. power served to imbue such ambiguity on its detention policy. The elements of ambiguity inherent in the U.S. War on Terror and the treatment of detainees in that effort were eventually challenged through their revelation in the Abu Ghraib photographs.

But let me back up for a moment to discuss the role of such opaqueness. As it did for many years in Vietnam in the form of a vague operational goal to "win," such ambiguity renders the practices of power possible because such practices are largely obscured and remain hidden from "public view." This is the basis for both their efficiency and their compatibility with a national power that prides itself on moralism. When images of discipline remain private, it allows democratic citizens to imag-

---

24. Greenberg and Dratel 2005, xviii.
25. Josh White, "Abu Ghraib Tactics Were First Used at Guantanamo," *Washington Post*, 14 July 2005.

ine them to be "not that bad."[26] Additionally, if we do not see the subjects of disciplinary practices, if they are concealed, then we can imagine (by "forming a mental image" of them) that they are as bad as possible. *The invisibility of the detainee is necessary for the citizen of the country detaining such individuals to perceive such treatment as necessary.* Readers familiar with her work will note here the overlap with Judith Butler's thesis on the "derealization" of those detained in the War on Terror.

> If violence is done against those who are unreal, then, from the perspective of violence, it fails to injure or negate those lives since those lives are already negated . . . They cannot be mourned because they are always already lost or, rather, never "were," and they must be killed, since they seem to live on, stubbornly, in this state of deadness . . . The derealization of the "other" means that it is neither alive nor dead, but interminably spectral. The infinite paranoia that imagines the war against terrorism as a war without end will be one that justifies itself endlessly in relation to the spectral infinity of its enemy.[27]

A popular assumption that permeated the United States during the initial years of the War on Terror—one that, to this day, continues to be invoked to justify torture done during those years—is the "ticking time bomb" scenario. This is a scenario in its fullest sense. The citizen has to imagine that such a scenario is being acted out in real time in order for them to accept that the harsh interrogation of detainees is both necessary and moral. Again we return to the Foucauldian and Lacanian assertions previously developed in chapter 1 and to those positions advanced earlier in the book regarding power's "imaginative" aesthetic qualities. It is because the detainees are hidden from public view that it was left to television programs, *forms of art and entertainment,* to fill the ontological vacuum of what the Other (terror suspects) and the Self (U.S. officials) were.[28]

This scenario has permeated U.S. society in its pop culture, specifically through the popular U.S. television program *24.*

---

26. "Photographic images of the treatment of detainees at Abu Ghraib invite the conclusion that public acceptance of torture is dependent on efforts to ensure that its concealment protects moral sensibilities" (Linklater 2007: 115).

27. Judith Butler, *Precarious Life: The Powers of Mourning and Violence* (London: Verso, 2004), 33–34.

28. James der Derian (2001; see also "War of the Networks," *Theory and Event* 5, no. 4 [2001]) has mapped out the "military-industrial-media-entertainment complex" (2002).

The show's [24's] appeal, however, lies less in its violence than in its giddily literal rendering of a classic thriller trope: the "ticking time bomb" plot. Each hour-long episode represents an hour in the life of the characters, and every minute that passes onscreen brings the United States a minute closer to doomsday. (Mayer 2007)[29]

According to Jane Mayer (2007), what is especially noteworthy is that "since September 11th, depictions of torture have become much more common on American television." In 24 alone, "sixty-seven torture scenes" have been depicted in "the first five seasons." While the torturers are sometimes the "villains" of the show, it nevertheless includes depictions of "heroic American officials" torturing suspects. "On 24," Mayer notes (2008: 196), "torture always worked." Further, Mayer notes that while the writer of the show had no experience in the military or in intelligence, one U.S. lawyer in Guatanamo admitted that Jack Bauer (the hero of the show) " 'gave people lots of ideas' as they sought for interrogation models that fall" of 2002 (ibid.).

An additional source to help "imagine" techniques of interrogation, according to Mayer, was the U.S. military's survival, evasion, resistance, and escape (SERE) program. This program, which dates back to at least the 1950s, trains military personnel on tactics they might use if they are ever captured or detained by enemy forces. To simulate such a capture scenario, the U.S. military prepares the would-be captured by "interrogating" them with coercive techniques (including waterboarding). This was the second model of interrogation put into action at Bagram Air Base in Afghanistan; Guantanamo Bay, Cuba; and Abu Ghraib prison in Iraq. So, to summarize, the models for coercive interrogation in the U.S. War on Terror came from (1) a U.S. television program and (2) a decades-old program that was meant to prevent soldiers from giving *false confessions* in the event that they were interrogated.[30]

How might these models begin to lose their efficacy? It proves more difficult than one would think at first blush. As illegitimate as they would seem to be to justify torture, the issue to note is that these models, while in the ether of the citizen's imagination, are nevertheless intuited as what is really "happening out there." To start, the aforementioned reflexive discourse form of counterpower can be a first step toward such

---

29. Mayer 2007.

30. "SERE is a training regimen—it's not designed to produce a truthful response" (Brittain Mallow quoted in Mayer 2008: 197).

"uncovering" of the unreal, to reveal those subjects to be less than that dystopic ideal of a monster or spectral—making them less unreal, as it were. There can also be a revelation of the Self by way of unfavorable historical analogy. An example of this comes from U.S. senator Richard Durbin, who has stated, in remarks made on the U.S. Senate floor on 14 June 2005, that the pictures of Abu Ghraib gave him a "sick feeling" (Mayer 2008: 259). After quoting from an FBI account that detailed the conditions of detainees' interview rooms and of the detainees themselves, Durbin stated,

> If I read this to you and did not tell you that it was an FBI agent describing what Americans had done to prisoners in their control, you would most certainly believe this must have been done by Nazis, Soviets in their gulags, or some mad regime—Pol Pot or others—that had no concern for human beings. Sadly, that is not the case. This was the action of Americans in the treatment of their prisoners.[31]

Here, the ontology of the U.S. Self is likened to the most immorally repugnant regimes in history. However, because Durbin only countered a description with another description, there still exists an opportunity for discourses of power to obfuscate the practice, the victimized subject, or both. Indeed, the reaction Durbin's remarks engendered in U.S. society demonstrate the disciplinary function of U.S. society that I have noted throughout this chapter, in that many considered Durbin's likening of the United States to these genocidal regimes to cross a line of propriety. The fact that Durbin *apologized* one week later for the remarks demonstrates reflexive discourse's limits in manipulating the internal and robust sense of the U.S. Self and the latter's power to discipline.

As Michael Williams has averred in his critique of the Copenhagen School's linguistic emphasis, experiences of publics are more likely and more powerfully constructed via images (2003: 526). Indeed, in this instance, it was the images of Abu Ghraib, rather than the discursive presentations, that engendered the ontological insecurity of American power's aesthetic Self. Javal Davis, a court-martialed Abu Ghraib military policeman, captured this reality of self-interrogation when he stated, "If there were no photos, there would be no Abu Ghraib."[32]

In October 2003, two individuals—Staff Sergeant Ivan Frederick II

---

31. http://durbin.senate.gov/gitmo.cfm.
32. http://www.hbo.com/docs/programs/ghostsofabughraib/synopsis.html.

and Specialist Charles Graner Jr., members of the 372nd Military Police Company—began capturing the abuse of Iraqi detainees at Abu Ghraib with a digital camera, with the first picture recorded on 17 October (Ricks 2007: 291). Again, while the *intent* of those who produce counterpower is not my concern here (indeed, Graner and Frederick, one would assume, would hardly intend the pictures to lead to an American aesthetic recoil), the question remains, why would these individuals record such abuses? While it is still unclear, one account suggests that following his time as a marine in the First Gulf War, Graner had returned to the United States and been denied benefits and psychiatric assistance from the Veterans Administration because the latter did not believe some of his combat stories. Thus, "this time around he said that he was going to have proof of whatever he saw and went through" (Lyndee England quoted in Gourevitch and Morris 2008: 134).

The person largely responsible for providing the photos to authorities was U.S. Army sergeant Joseph Darby. Having come across the pictures in January 2004, Darby placed the compact discs on which the pictures were saved into an envelope for the U.S. Army Criminal Investigative Command. As it eventually did with the My Lai massacre, the army launched an investigation, headed by army general Antonio Taguba, into the abuses. The pictures were first made public on the U.S. news program *60 Minutes II* on 28 April 2004. The poses included one where a detainee is placed on a wooden block, in a hood with alligator clips attached. Another series of photographs showed naked Iraqis in humiliating poses. Following the CBS program, Seymour Hersh—the same reporter who helped uncover the My Lai atrocities—published the first detailed written report over the abuses, in a series of articles in the *New Yorker* magazine throughout May 2004. Of course, the photos themselves generated a flurry of discussion over who or what was to blame for the abuses, but I find several particular responses noteworthy. The first of these came from Senator John McCain, who claimed on 6 May 2004 that "decisive action" was needed to raze the prison itself because it had become a symbol of "torture and mistreatment." Here the architecture of discipline is the target of power—rather than the aesthetics of the painter or the painting.

The second noteworthy issue is the references throughout about the "sickening" feeling Americans recorded when they viewed the photographs. For Americans, "the body's [U.S. power's] defects, together with the stains and the mess [Abu Ghraib], will be imprinted on the [American] mind, giving rise to disgust" (Foucault 1984: 138). This perhaps helps explicate the meaning behind U.S. president George W. Bush's ironic remark that Abu Ghraib was "a *stain* on our country's honor and our country's reputation . . . Americans, like me, didn't appreciate what we saw, that it made us *sick to our stomachs*."[33] Ironically, the policies that resulted in this treatment were instituted by Bush's own Justice Department. Yet the remark is also telling because it implies that while such coercive interrogation policies were advanced by the Bush administration, the shock of the *image* of Abu Ghraib, once such an image was publicized, was a spontaneous event, generating aesthetic insecurity by challenging the beauty of the "idea" that is U.S. power.

Foucault posited that torture as a "spectacle" was on the decline because of the internalized self-discipline that accompanied modern forms of disciplinary surveillance.[34] From this perspective, the explicit U.S. institutionalization of detainee abuse does not represent a further spread of U.S. hegemony but signals, instead, a decline in U.S. influence and ability to discipline—a decline further accelerated by the resentment produced by such images of U.S. abuse of Arabs. But the vulnerability of a power that commits these acts to an undeterminable counterpower is even more pronounced considering three further observations. First, as mentioned in the first chapter of this book, advances in technology—often coinciding with generations—have made the gathering of the image, of "data," at once both micro and surreal. The image can now be downloaded onto flash drives (or burned onto a CD, as it was in the case of Joseph Darby's gathering of the Abu Ghraib photos), but they can then be "uploaded" into online storage cottages—cyberspace. Such images, data, and accounts elide control. They can never be destroyed, and thus their unreality is their most potent constitutive element.[35]

A second observation is that while the imaginative nature of the "ticking time bomb" scenario bestows power with the ability to continue such coercion, such coercion also contaminates the aggregate amount of information being used to secure the Self from future "unimaginable" attacks. In this case, the vitalist condition of power's need to act is

---

33. "Bush Apologizes, Calls Abuse 'Stain' on Nation," *Washington Post*, 7 May 2004.

34. "At the beginning of the nineteenth century, then, the great spectacle of physical punishment disappeared; the tortured body was avoided; the theatrical representation of pain was excluded from punishment" (Foucault 1977b: 14).

35. Of course, the long-term result of such a development could be that data swamps us. With too much information floating around, the need to uncover through an archaeological dig becomes more difficult, as more layers cover up the counterpowerful image more quickly. Nevertheless, the image remains—somewhere—that can challenge power at a "moment's notice."

magnified. It must move, do something, "demonstrate" its physique, foil more attacks. Take for example the account that Ron Suskind (2006: 115) relates regarding the treatment of Abu Zubaydah. Zubaydah is a terrorist suspect and detainee who was waterboarded some eighty-three times by interrogators (Bradbury 2005), beaten, and told of "his impending death." Under such conditions, Zubaydah told his CIA interrogators that shopping malls, banks, supermarkets, water systems, and even nuclear plants were targets of al-Qaeda plots. Suskind writes, "Thousands of uniformed men and women raced in a panic to each flavor of target . . . Where would they start sniffing? No idea. Start with . . . *everywhere*" (2006: 115–16).

Tellingly, it was this series of interrogations and specifically the false information that resulted from them that President Bush, in a speech in September 2006, used to justify his administration's entire interrogation policy (Mayer 2008: 155). But what Suskind's account suggests is that against an enemy that cannot be fully determined, in a war on terror, the ambiguity that allows power to operate "in the dark" produces *bad information*—power's interrogative actions produce more actions. This is precisely where a counterpower understanding of the aesthetic problems that, when revealed by interrogative images, produce a recoil of power to the "ugliness" of the Self link and intersect with more utilitarian arguments against torture (see Arrigo 2004).

A related, third observation is that the way a threat is framed makes power inherently vulnerable. The ambiguity of the Other makes a war on terrorism necessarily vague and amorphous and difficult to calculate and operationalize. As one author notes, "the 'global' war started by the United States against terrorism is a war between rootless, deterritorialized partisans, who are organized essentially as networks" (de Benoist 2007: 78). These observations illuminate one of the central assertions of this book—that aesthetic power facilitates its own challenges.

We can further explicate some of the aesthetic implications of these abuses by first focusing on the sexualized and gendered nature of U.S. detainee abuse. The argument here focuses on, for instance, the images of a female interrogator such as Lyndee England pointing to the groin area of a naked Iraqi. By confronting Arab sexuality, the United States did more than demonstrate authority; it shattered the Arab sense of Self.[36] The Bush administration and its intellectual supporters have in-

---

36. Phillippa Winkler notes that "the combination of sexual humiliation and physical abuse combines to destroy cultural and national cohesiveness" ("Sexualized Torture by US Troops: Implications for the North South Divide," paper presented at the annual meeting of the International Studies Association, San Diego, CA, March 2006, 11).

deed justified the humiliation (if not sexual in nature) of Iraqis as a necessary tool used against a culture where such shaming resonates. Deputy defense secretary Paul Wolfowitz asserted in April 2003 how the "images of people pulling down a statue and celebrating the arrival of American troops is having a shaming effect throughout the region."[37] Furthermore, as Seymour Hersh reported in May 2004 (shortly after breaking the Abu Ghraib story), "the notion that Arabs are particularly vulnerable to sexual humiliation became a talking point among pro-war Washington conservatives in the months before the March, 2003, invasion of Iraq." Hersh stated that many of these individuals had read and used Raphael Patai's book *The Arab Mind* as its theoretical resource: "The Patai book, an academic told me, was 'the bible of the neocons on Arab behavior.' In their discussions, he said, two themes emerged—'one, that Arabs only understand force and, two, that the biggest weakness of Arabs is shame and humiliation.'"[38] By exposing and challenging Arab sexuality(through engaging the taboo of homosexuality, for instance), the treatment of prisoners at Abu Ghraib was a further extension of U.S. authority and, ostensibly, U.S. security practices intending to demolish the Arab sense of Self, as evidenced by the photo taken at Abu Ghraib of a "pyramid" of naked Iraqi men.

The Nausea That Is No Longer a Passing Fit?

Who do we "blame" for the production of these abuses? This question is particularly relevant for ongoing debates in the United States regarding torture. In exploring an answer, we must also peg the mechanisms responsible for the abuses, and such mechanisms exist through the interactions of individuals in the production of U.S. aesthetic power. Such production will inevitably create these abuses when operating "in the dark." A key claim I made in a study preliminary to the current one (Steele 2008b) was that once these abuses were "revealed" in photos, the revulsion to such abuses by the gaze of the U.S. Self would be enough to at least force a rethinking (in terms of political debate). I concluded that study, however, by arguing (via Sartre 1938: 170) that once this revulsion (i.e., "nausea") is no longer a "passing fit" but "is I [United States]," then torture will have lost its moral repugnance for a collective U.S. Self that has come to accept it as part of the political

---

37. Bill Gertz and Roman Scarborough, "Wolfowitz Sees 'Shaming Effect' on Arab World," *Washington Times*, 29 April 2003.

38. Hersh 2007.

body's daily functioning. In the closing of this subsection, I address that possibility. More immediately, when seeking "blame" for the abuses, some, of course, exists with those individual members of the military who committed the acts. Yet in light of recent surveys that have shown, albeit in a varied manner, rising support by Americans for torture (Pew 2008, 2009), this indicates that such abuses *continue* to be implicated in a deeper disciplinary process of the U.S. Self's subjectivity. I share Karen Greenberg's assertion that "these concerns [over such practices] are the beginning of a debate which must ultimately call into question not the Bush administration but the American people. The use of coercive interrogation techniques was downplayed, not only by the military, but by the US press as well."[39] Several examples are illustrative in this respect and lead us to consider whether uncovering images of such abuses in the future will contain the same ontological shock as they did with Abu Ghraib.

One example of the Self's disciplinary power is evident through the experience of Joseph Darby, the whistle-blower of Abu Ghraib, who could not return to his home in Cumberland, Maryland, because of the threats to his life that he faced from fellow Americans who considered him a "rat," "traitor," and/or "someone who let his unit down."[40] Darby is being disciplined not just for betraying the loyalty of his fellow soldiers but for providing the pictures that stain the aesthetic integrity of U.S. power, for disconnecting the emotional link between the U.S. Self and its citizens. Discipline was also at work on 28 September 2006, when the U.S. Senate passed the Military Commissions Act of 2006 *by nearly a two-to-one margin*.[41] The act included provisions that barred terrorist suspects from their habeas right to challenge their detentions in court. While outlawing certain explicit interrogation practices, the act also allowed the executive to independently define further permissible interrogation techniques. Yet another example of the disciplinary Self was revealed in a debate among the presidential candidates for the 2008 Republican nomination, where one candidate appeared to support the use of waterboarding while another supported the "doubling" of the U.S. detention

---

39. Greenberg and Dratel 2005, xviii.

40. "Exposing the Truth of Abu Ghraib," http://www.cbsnews.com/stories/2006/12/07/60minutes/main2238188_page3.shtml.

41. The vote tally was sixty-five to thirty-four. See http://www.senate.gov/legislative/LIS/roll_call_lists/roll_call_vote_cfm.cfm?congress=109&session=2&vote=00259#top.

center in Guantanamo Bay, Cuba. What is most revealing is that, according to one account, these answers "drew *enthusiastic applause* from the Republican audience."[42]

I do indeed see as a *positive resource* the national "shame" that many Americans experienced in 2004 when such detainee abuses were uncovered, in that it is "central to safeguarding the freedom of the body, hence, to keep alive our freedom to act responsibly."[43] As Linklater notes (2007: 117), "countervailing normative power is found nonetheless in the belief that shame is attached to the 'sacrifices of value' which have occurred in the 'war against terror.'"

Yet while the imaginative stratum for aesthetic power explicates why its operation "in the dark" leads to problematic outcomes, it also proves insightful for why certain populations of Americans, including those who are most religious, express a majority level of support for torture in recent polls.[44] While there is some slight variation depending on denomination (white evangelical Protestants are most supportive of torture, white mainline Protestants are least supportive, and white non-Hispanic Catholics fall somewhere in between), a clear relationship comes through in one particularly vibrant survey: those who attend religious services most frequently are most supportive of torture. "Unaffiliated" individuals and those who attend religious services "seldom" or "never" are least supportive of torture. That Christians would support torture provides a stark juxtaposition, of course, and a snap judgment of such polling could be that this is what happens when both torture and religion are politicized in a democratic polity. Yet let us return to the imaginative implications that might be explicated from these survey results.

First, if the mechanism for such support is the intensity of "beliefness" in a population, which exemplifies the obstacles (and also the opportunities) that counterpower faces, such support for torture makes sense (even if it is ethically problematic). Those who see their "faith" as particularly strong—who are willing to "believe" that something exists that they can imagine but never see—are those who will be more willing

---

42. Peter Wallsten, "GOP Candidates Divided on Detainees," *Los Angeles Times,* 17 May 2007, 15. Republican presidential candidate John McCain opposed this view.

43. Jean Bethke Elshtain, "Shame and Public Life," in *Trauma and Self,* ed. Charles B. Strozier and Michael Flynn (London: Rowman and Littlefield, 1996), 285.

44. "Polls Show Support for Torture among Southern Evangelicals," http://pewforum.org/news/rss.php?NewsID=16465; "The Religious Dimensions of the Torture Debate," http://pewforum.org/docs/?DocID=156.

to imagine that the torture of individuals is both necessary and unproblematic. Second, such faith also lends itself to a belief in the righteous intent of authority, a trust that those who operate in the "shadows" are doing so for reasons of physical security. Yet even more recently, the support for torture or at least for indefinite detention extends to even wider sections of the American electorate, with almost two-to-one majorities opposed to the closing of the key detention center of Guatanamo Bay.[45]

These recent developments, then, require us to ask whether the nausea that Bush and Durbin recorded feeling may no longer be a "passing fit," in the words of Sartre, but may be "I [America]." Will an uncovering of future or current cases of abuse repulse the gaze of U.S. power? Will the latter even view the former as a "de-aestheticized" picture? There are two possible answers to this question. One possibility, evidenced by the revelation that the U.S. Central Intelligence Agency destroyed ninety-two tapes of interrogations it committed, is that there still exists counterpower potential in photos.[46] Yet it seems that for many of the examples I have provided here regarding the "disciplinary" elements of the U.S. Self, what was once an "ugly" representation of the U.S. Self may no longer be. Francois Debrix's contention that agonal action itself *is beauty*—that even violence, properly expressing "passion, emotion, and heroism," becomes beautiful and "has no value or essence on its own, other than that which is represented and aestheticized every time the heroic agonal agents . . . perform their agonal and violent deeds" (2007: 114–15)—may have, in the instance of the U.S. practice of torture, come to pass.

Conclusions

Technological advances have always served to destabilize centralized power—that is, until such power or rivals to power become able to master these advances to further their own control. We may, however, be at a point in history where power has met its match. Twenty-first-century

---

45. Several polls were taken in the spring of 2009 when the Obama administration announced that it was planning to close the detention center. The results of those surveys can be viewed at http://pollingreport.com/terror.htm.

46. "CIA Destroyed Interrogation Tapes," http://www.telegraph.co.uk/news/world news/northamerica/usa/4929026/CIA-destroyed-terror-interrogation-tapes.html.

power is now at a point where it cannot corral the speed and vibrancy of the image. As I already mentioned, advances in technology have made the gathering of the image, of "data," at once both micro and surreal.[47] This creates a condition where power is perpetually suspended because it cannot know when an image of its counteraesthetics will emerge. In essence, counterpower has been *democratized*—anyone can disseminate the unfavorable image that unfastens power. I do not see this as a development that implicates just U.S. power or even the power of states, even though it is regarding the former that I have here drawn examples for illustration. One final example demonstrates how civil society organizations have used the speed provided by (post)modern technology to destabilize the control of power.

In 2007, B'Tselem, an Israeli human rights group, began handing out video cameras to Palestinians living near Jewish settlements, in order to capture any attacks that the Jewish settlers might execute on the former, calling it the "Shooting Back" project. In 2008, a group of Palestinians recorded such an attack in the West Bank.[48] The video shows a group of masked men armed with baseball bats—presumed to be Jewish settlers—beating an elderly Palestinian couple. The video is only a little more than a minute long, and the beating itself is difficult to see (as the woman filming drops the camera almost immediately after the beating commences, to come to the aid of the victims). But the images are striking, and the sounds of the attack are audible. As of this writing, the investigation into these beatings continues, but what is important is the B'Tselem director's recognition of the Shooting Back project's organic import: "When they have the camera, they have proof that something happened. They now have something they can work with, to use as a weapon."[49] The instantaneity of such counterpower and the fact that no one knows who has (or does not have) one of these provided cameras serve as a Panopticon reversed back against violent power—it does not know who or what is watching.

Again, however, we must wait and see, for I am not ready to suggest once and for all that we are at the "end of history" when it comes to tech-

47. This is why I am skeptical that the aforementioned CIA tapes recording the torture of terrorist suspects are ever completely "destroyed."

48. The video is embedded in an online BBC account of the attack at http://news .bbc.co.uk/2/hi/middle_east/7451691.stm.

49. Oren Yakobovich quoted in "Jewish Settler Attack on Film," http://news.bbc.co .uk/2/hi/middle_east/7451691.stm.

nology's ability to permanently suspend aesthetic power. Future research should probe this possibility more thoroughly. In the next chapter, based on the theoretical arguments advanced in the introduction and chapter 1 and on the analytical insights illustrated by empirical examples in this chapter and in chapters 2 and 3, we can now situate aesthetic power and counterpower among other ontologies of power in IR theory.

FIVE

Toward a Transgressional Account of Power

To this point, this book has advanced the notion that an aesthetic insecurity operates in centralized bodies of power and that devices titled "counterpower" can engage this aesthetic insecurity. Going forward, however, one might ask whether counterpower can be situated within a larger body of power analysis. I argue here that it can—casting counterpower as a form of transgressional analysis. To do this, the present chapter contains several moves. First, rather than critique or even comprehensively review any and all literature on power, it takes a more modest route in demarcating two other categories where power contestation can be analyzed. Second, in terms of temporal and spatial frames, this chapter asserts how the transgressional analyzes the "instantaneous play" of power's vulnerabilities as they appear within seconds, minutes, and days (Foucault 1977b: 37). In the transgression, a subject perceives itself to be autonomous, yet through this autonomy, it facilitates a limit that is breached. Unlike the spaces and periods in transcendental and transitional orders (where power has been contested or upended through identifiable rivals), in the transgression, the Self (which is always ambiguous) immediately is "lost in this space it marks with its sovereignty and becomes silent" (35). The transgressional perspective, because it analyzes at this "busier" level of contestation, needs to also be understood less as a theoretical school and more as a philosophy that emphasizes creation and emergence.

Finally, this chapter demonstrates how the turn to aesthetics and power that has been the basis for this book is a logical one considering the ethos of creation embedded in the transgression, and it suggests that future analyses that focus on the creative forces that spring forth in global politics might benefit from embracing transgressional analysis. To do so, such studies would need to jettison the notion of an unproblematized subject. For the transgressional level, if we take as a given that there is no grounding or origin for the Self, then it makes little sense to maintain a scientific posture directed toward its analysis. To summarize, I argue in this chapter that because the bulk of IR analyses that have examined power to date have looked for its operation in certain locations and within periods of time, which has made possible the explication of such operation as patterned, there has been little need to use philosophy to develop their analyses. The transgression, by contrast, eludes probabilistic specification and thus represents more of a philosophical ethos than a predictive theory. Nevertheless, intersections between transcendental and transgressional analyses, on the one hand, and transgressional analysis, on the other, are important enough for us to view the latter less as a critique of the former two and more as a complementary mode of viewing the subject.

This chapter begins with a brief generalization of transcendental and transitional analyses of power, with some examples, and explicates the temporal and spatial conditions, as well as the function of the intellectual, in each of these traditions. It then contrasts these with the temporality and spatiality found in a transgressional analysis and explicates the trangressional insights found in the work of Andrew Bacevich (2005) on "the new American militarism." For final illustration of these three traditions of power analysis, the chapter concludes with the 2004 Fallujah bridge case, which was analyzed in the previous chapter.

Transitional and Transcendental Analysis of Power

The majority of accounts dealing with power in international relations have, to date, been *transitional*. Transitional accounts of power conceptualize it as relational in form. To derive influence from power presumes that there is a community that recognizes what is powerful and interacts according to this logic. Max Weber, for example, defined power as the "opportunity to have one's will prevail within a social relationship, also against resistance, no matter what this opportunity is based on" (Weber 1956, as translated in Berenskoetter 2007: 3).[1] The role of the Other in

---

1. Uphoff (1989: 299) takes issue with this general tendency—exemplified here by Berenskoetter but going back to Simon (1953) and Peter Blau (1964), among others—to "affirm that power *is not a thing but rather a relationship.*"

the transitional landscape is important, for it is where subjective notions of power intersect that we have the intersubjectivity in its meaning, which I discuss in more detail in the following section.

We can therefore organize a surprisingly wide array of perspectives as transitional. Such a logic permeates recent constructivist undertakings of power as a form of "moral authority" (Hall 1997) and "moral prestige" (Löwenheim 2003), Habermasian IR perspectives (Cruz 2000), and models that presume what Janice Bially Mattern titles "communicative exchange," defined as "the process through which actors convey their interpretations and perceptions of things in the world to each other . . . mediated through language . . . it enables actors to transform their subjective, privately held opinions about what they see in the world into public information" (2007: 106–7). As Bially Mattern also notes (101), Nye's now famous concept of "soft power" (Nye 2004) rests on a logic of communicative exchange—as political ideals and values of a national subject become attractive when communicated and legitimated into and through an international society.

The patterns of power balancing (found in realist accounts) and international societal shifts (found in English School perspectives) are also transitional and based on relational logic.[2] A primary example comes from the important neorealist work of Kenneth Waltz (1979, 1988). While it is based on a microfoundation, it also assumes a logic where systemic distributions of power determine and shape the behavior of state units. The international community recognizes systemic or societal imbalances and, having an interest in homeostasis, seeks to rectify those by "balancing" or, in the case of an international society, by privileging a constitutive principle and reinforcing it because it provides international order. For realists, this occurs through a recognized imbalance of power that the nation-state reacts against. For the English School theorist, balancing is a societal good to be upheld to ensure order and predictability.[3]

Power transition theory, which proposes long cycles where different national actors serve as a dominant systemic power, would seem to contain very little relational or intersubjective logic, as it defines power in

2. Richard Little (2007) has distinguished this balance-of-power binary as encompassing either an associational or adversarial logic (esp. 66–68), sometimes at the same time (see especially his "proto-constructivist" interpretation of Morgenthau in chapter 4).

3. Little posits that in Hedley Bull's account, the balance-of-power "concept [itself] has no meaning in the context of an international system because it is an institution that requires the existence of a common language and shared beliefs" (2007: 143).

concrete hierarchical terms[4] and sees "cycles" occurring in patterns predictable and measurable through time. Yet even here, we see a relational
logic. The key variable that determines when a rising power will "over-
take" a dominant one is satisfaction or dissatisfaction, or, more precisely,
"relative satisfaction with the rules of the global or regional hierarchy"
(Tammen et al. 2000: 9). Thus, the durability of the legitimacy of the
rules a dominant power institutes determines whether and when a transition can occur and, if it does, whether it will occur violently.

Unipolar and hegemonic stability theories are also transitional. This
may prove to be confusing at first blush, as these perspectives assume
some hierarchy, rather than anarchy, obtaining in certain epochs. Thus,
power here appears to be used and then becomes vulnerable in a transcendental sense. Imperial accounts (Nexon and Wright 2007) view
power in what I will discuss shortly as transcendental terms, whereas
hegemonic stability theory (which includes power transition theory) and
unipolarity are transitional accounts of power.[5] While imperial orders
are transcendental because they are explicitly hierarchical and diffuse,
unipolar and hegemonic orders are transitional because authority, while
centralized, is still relational.[6]

The temporal frame in transitional accounts of power, while not as
comprehensive as the transcendental frame, is still substantive. Hegemonic power transitions occur over "long cycles," and thus hegemonic
wars occur at the pattern of once every fifty (Goldstein 1988) to one
hundred years (Toynbee 1954; Modelski 1978; Gilpin 1981; Boswell and
Sweat 1991). Other shifts in the distribution of power, say from multipolarity to bipolarity, can occur somewhat more frequently, for as the number of autonomous "poles" in a system increases, the possibilities for
changes in national power increase as well. These do, however, happen
over a matter of at least a decade, as the post–Cold War transformation

---

4. "Power is a combination of three elements: the number of people who can work
and fight, their economic productivity, and the effectiveness of the political system in ex-
tracting and pooling individual contribution to advance national goals" (Tammen et al.
2000: 8).

5. Imperial orders include "network properties" such as "*rule through intermediaries* and
heterogeneous contracting between imperial cores and constituent political communities"
(Nexon and Wright 2007: 253), whereas no such conditions obtain in unipolarity or hegemonic orders.

6. From a more interpretive perspective, Richard Ned Lebow turns to the Greeks, who
used words for hegemonic rule (*hegemonia*) as a form of legitimate authority—recognized
by the community because of the benefits it accrues from the hegemon. *Arche*, "by contrast,
means 'control'" (Lebow 2007: 124) and need not require any consent from subjects.

seems to have taken. The emergence of new principles constituting international society also takes time, as seen through Hedley Bull's analysis (1977: chap. 2) of Christian society (fifteenth through seventeenth centuries), European society (eighteenth through nineteenth centuries), and world international society (twentieth century). The analysis of whether international society is transforming into a world society (see Clark 2007; Buzan 2004) and thus whether the transitional can be transcended by the global also includes a time frame of at least one hundred years ("beginning of the 19th century," according to Clark 2007).

Transcendental analyses of global politics assume that the order of international politics not only is subject to change but can be upended and reformed so that a global consciousness, either imposed or dialectically manifested through human group interaction and struggle, develops in a way that anarchy can be obviated or even replaced. Transcendental insecurity manifests when the existing order of a global system is delegitimized or challenged and an alternative order arises to take its place. Intellectuals also play a role, to varying degrees, in helping to imbue legitimacy to the transcendental order. Transcendental perspectives include those termed cosmopolitan (Shapcott 2001), Kantian (both in a confederative or consistent forms), or neo-Gramscian (Cox 1983; Gill 1990; Gill and Law 1988); the so-called Stanford School or world polity theorists (Meyer et al. 1997; Boli and Thomas 1999); and the recent constructions of a global (Shaw 2001) or world (Wendt 2003) state.

Neo-Gramscian accounts are transcendental as they serve to depict hegemony as not only being challenged but overturned—ideologically and materially, albeit on a very infrequent basis. These approaches see an important role in maintaining transcendental order for civil society organizations. Such associative groups—whether at the domestic level, with churches, an educational system, and a national press (Cox 1983: 164), or at the international level—are supposed to function as outlets against the prevailing transcendental order. Somewhat counterintuitively, Cox argues, these potential sources for counterhegemonic ideas are instead co-opted by successful hegemons and even serve to "complement and support" hegemonic goals for the world economy (174).

In the first few years of the past decade, Martin Shaw (2000, 2001) and Alexander Wendt (2003) asserted that the world was moving toward a form of a global sovereign. Shaw's argument proceeds through four assumptions. First, global actors share a common struggle derived from negative experiences. Much like the substrate out of which Michael Ignatieff could construct his liberal conception of human rights, where the common memory of the Holocaust could "lay bare what the world looked like when pure tyranny was given free rein to exploit natural cruelty" (2000: 81), Shaw proceeds from the observation that common negative experiences ignite the struggle. From this struggle, secondly, there has arisen a global set of values or collective consciousness. These values then, thirdly, constitute international and global institutions, which help protect and reinforce those values. Fourth, this has led, in the wake of the Cold War, to democratic nation-states all working to foster a global democratic state (Shaw 2001: 638). Power, in both a material and authoritative form, constellates in this global democratic state, one that is futilely resisted by nondemocratic entities.

Alexander Wendt's argument about a world state is similar, although it departs from Shaw in that it sees Shaw's global state as inconsiderate of several possibilities, including that of a "rogue Great Power" (Wendt 2003: 506) that acts against the community. Here again we see the eventual obviation of the existing order, as "the past [for a world state] is anarchy" (503). Wendt's teleological argument utilizes several causational processes. First is the upward causation of the microlevel, where individual decisions lead to patterns of behavior that work back on units in the form of structures. The second process, downward causation, "refers to the way in which boundary conditions constrain and govern the interaction of a system's parts" (500). It is in the third process, "final causation," where we find power existing (Drulak 2006: 155), as the struggle for recognition provides for instability and as an asymmetry of identities (unequal recognition) "pushes" the system toward an end state. But because the "super-macro-structure" of the world values stability, *power is used by the system* to inhibit local attractors by encoding and revising boundary conditions to constrain threats to the system even further (Wendt 2003: 502).

Transcendental orders are developed and challenged over the course of centuries or even millennia. As was discussed earlier in this book, certain processes analyzed by Neta Crawford in her thesis on the role of ethical argument in promoting change, such as decolonization and the abolition of slavery, included a transcendental account, with change occurring "over the course of several hundred years" (2002: 133).[7] More comprehensively, Wendt (2003: 503) frames the development of the

---

7. Other empirical phenomena analyzed by Crawford included a transitional logic, such as humanitarian intervention (2002: 399–434). I engaged Crawford's mode of ethical argumentation analysis explicitly in chapter 2 of the present study, where it is contrasted to the "reflexive discourse" form of counterpower.

world state by contrasting the number of political communities in existence three thousand years ago (six hundred thousand) to the number today (two hundred). Robert Cox's analysis of global hegemony begins with the historically contingent period of the early nineteenth century, where social conflicts in domestic spheres transferred into a global environment where international financial concerns could, by the early twentieth century, serve "to reconcile social pressures with the requirements of a world economy" (1986: 230).

One of the unique functions that makes transcendental orders slow to change stems from the practice of intellectuals in imbuing such orders as "natural" or unproblematic, whereby they function within regimes of power to help "frame thought and thereby circumscribe action" (Cox 1992: 179). While the world may be institutionalized along national sovereign boundaries, the intellectual community *exists transnationally*. Part of the solidarity that is propagated and reinforced in a "global state" emerges from the bonds existing across this community, as exemplified by Martin Shaw's nod to Tehran to observe "our counterparts in universities there" (2001: 632). Members of the world polity school argue that professional associations of a "rationalized world institutional and cultural order" develop models for world development and, thus, that world society impacts nation-states toward isomorphism, in part, because of these professional networks (Meyer et al. 1997).[8] One can only assume that such a community will continue to develop in the information age, where high-speed Internet and airline travel allow for rapid exchanges of intellectuals and their ideas.[9]

Bruce Russett, an individual who embraces the tenets of explanatory social science, nevertheless also promotes the notion of the intellectual energizing the change toward the democratic peace proponent's preferred transcendental order, to create a "self-fulfilling prophecy," as I have discussed elsewhere (Steele 2007a). Russett counseled scholars to go about

---

8. This is hardly a new observation, however, as Peter Haas's (1989) seminal article on "epistemic communities" demonstrates the coalescing function that intellectuals provide, cutting through national cleavages by synthesizing scientific knowledge among national units.

9. Of course, it is not clear, by any means, that intellectuals would consider it their calling to participate in the type of transcendental change that Shaw thinks they should be participating in. Not all scholars in the IR field, for that matter, consider the production of knowledge to be oblivious to national boundaries, as studies by Ole Wæver (1998) and Steve Smith (2002) have identified.

> repeating the norms [of democratic peace] as descriptive principles
> [in order to] help to make them true. Repeating the proposition that
> democracies should not fight each other [in order to] reinforce the
> probability that democracies will not fight each other. (1993: 136)

Intellectuals also serve to paint transcendental orders as legitimate
through the institution of "developing and sustaining the mental images, technologies and organizations" (Cox 1983: 168). They thus not
only provide the content of the values of this order but regiment those
values so that the order appears normal and universal—the product and
possession of not "one class" but entire humanity (ibid.). They can also
serve to smooth out its rough edges—solving the problems within a
technoscientific framework that serves to uphold the power it places
there (Cox 1986). Such "doxic" knowledge goes "unchallenged and
unreflected upon" (Ish-Shalom 2006: 571).

In transcendental contestation, the intellectual can serve to challenge the status quo by doing what critical scholars have done all along—
making strange that which appears to be the normalcy of that order.
However, the intellectual cannot stop here, for according to Shaw, the
point is not only to explain or understand the world—"the point is to
change it" (2001: 640). Shaw also suggests that many of the democratic
movements in the "global democratic revolution" of the 1990s were led
by intellectuals and their students and that, in a more sobering respect,
these were, in no coincidence, the primary targets of totalitarian and reactionary forces seeking to gain a firmer grip on their loosening reins of
power.

In neo-Gramscian accounts, intellectuals may also infiltrate the all-im-
portant civil society sphere, where they can assist such a sphere to "act towards an alternative social order at local, regional, and global levels"
(Cox 1999: 28). More broadly and more expansively, Cox and other
neo-Gramscians have identified the role of intellectuals both in the legitimating sphere (entrenching power) and in a war of positioning
where a historic bloc must develop to challenge the existing transcendental order. What is necessary for a successful global revolution is the
buildup of these counterhegemonic forces. Wars of positions might perhaps precede, even obviate, the deployment of violence through wars of
movement. They title this "counterhegemony," which would eventually
supplant not only the physical presence of power but, more important,
the ways in which humans "conceived of the limits of the possible in their

own lives, as well as seeing the potential for a new type of society" (Gill and Law 1988: 64).

Yet to promote a counterhegemonic order proves difficult for intellectuals because such an order has been imbued as "universal in form," as not being contingent on class or nation.[10] Intellectuals who participate in establishing the rigidity and certainty of "laws" of social science, for example, craft further certainty and rigidity in the relations of social beings. Thus, the "trenches" of struggle are in the arena of knowledge production itself, yet even if a counterhegemonic force emerges in this struggle, there is no guarantee that it will not itself become invested by power for the purposes of creating greater control of, rather than greater autonomy for, human beings (Diez and Steans 2005: 129).

The Space between Transcendence and Transition

Spatial differentiation simply refers to the locale in which power operates, or, more precisely, as Giddens explains, the "setting of interaction . . . Locales may range from a room in a house, a street corner, the shop floor of a factory, towns and cities, to the territorially demarcated areas occupied by nation-states" (Giddens 1984: 118–19). A helpful starting point to contrast the logic of power through space is Arnold Wolfers's categorization of national "goals" (1962: 72–78), one that has been referenced by a swath of studies and commentaries since it appeared.[11] Possession goals are those used to "enhance the preservation of one or more of the things to which it [a nation] attaches value" (ibid., 72), whereas milieu goals are, as the name suggests, those intended to transform the environment within which an actor operates. Possession goals are adaptive and reactive—an actor exists within a particular environment that is held largely constant. Milieu goals are transformative and proactive—they are transcendental.

Several studies have viewed this difference in terms of the operation

---

10. However, in the long run, by claiming the relations between these social agents as "law-like"—when proclaiming, for example, that certain theories are "empirical laws" (Levy 1988) or a "fact" (Russett 1993)—they are serving to set up such relations for failure, for uncertainty, when they are compromised.

11. See especially its invocation by neoconservative commentator Joshua Muravchik (1992: chap. 3) and scholars such as Stephen Walker (1978), Stanley Hoffmann (1996), and Robert Keohane (2000: 128). Wolfers's categorization of goals has also found recent use in studies seeking to explicate the future for transatlantic relations between the United States and the European Union (see John Peterson 2004).

of international actors in between transcendental and transitional orders, where there exists both a milieu and a possessive nature to the goals of particular actors. Many center on the difficulty of changing the milieu of global politics—how anarchy rejects such change. Even so, such a space between transcendence and transition proves to be a fascinating terrain to investigate. John Herz's (1950) seminal thesis on the constraints that various forms of "idealist internationalism" have faced once entering the anarchical order provided one of the first explications on the difficulty in changing a milieu. For Herz, many movements over the past three centuries, spanning from the French Revolution through the Bolshevik Revolution to Wilsonianism, idealized themselves as world revolutionary (milieu-establishing) but were all inevitably disciplined by the existing milieu's rules of power politics.

In many cases, other members of a milieu coordinate a response when the milieu's status quo has been challenged. A recent study by Barak Mendelsohn (2005) demonstrated how al-Qaeda has been viewed as a societal threat because of its intended aims at an alternative political community—the caliphate—as opposed to the secular sovereign state. Judged in these terms, this attempt at changing the milieu has led to a reaction by great powers in eradicating a "parasite" of global politics (Löwenheim 2006). Thus, power here has an *authoritative-constitutive* function—it determines what constitutes a legitimate or functional member of the global sphere and what does not, and it wields violence in certain contexts to enforce this membership. Rodney Hall titles this "deontic power," which serves to "principally empower some actors, and not others. Deontic powers, in the social realm, create social relations of power, such as rights, duties and obligations" (2008: 37).[12]

Another example where we see a milieu transformation stymied comes in the form of an aspirant power that ignores the (transitional) principles of an international society, principles that served to stabilize an international order.[13] Such a "revisionist" state may attempt to transcend these principles for many purposes—the creation of a more just order, the acquisition of more territory for added security, or simply to

---

12. By contrast, an *allocative* function of power determines those resources that provide the materials of an environment and the means of production (see Giddens 1984: 258–59).

13. Paul Schroeder identifies this as one of the strategies available to states as an alternative to self-help. It was "a strategy less common, but far from unusual or unknown," which he titles "*transcending*, ie, attempting to surmount international anarchy and go beyond the normal limits of conflictual politics" (1994: 430, emphasis added).

satisfy an unsatiable hunger for more power. Regardless of the reason, such a nation-state will pay a price, as the community that still finds these principles worthy of protection will balance against such an actor. In the words of George Liska, "the aspirant to hegemony who defies the system and seeks release from its rules and restraints, imperatives and inhibitions, is the tragic hero in interstate relations" (1990: 190). For evidence here, we might point to most of the great powers' *hiding* behavior during the U.S. preventive war in Iraq.[14] Here, other great powers prevent an international societal milieu from being transformed into an imperial system.[15]

Finally, contemporary debates on "international" and "world" society, as evidenced in the recent wave of English School studies, are also, in many ways, centered on this "space between" transitional and transcendental orders, whether one concludes that international society is progressing toward something that resembles the analytical "ideal type" of world society (Buzan 2004) or whether, as Ian Clark (2007) suggests, we should construct world society as a historical referent that has existed alongside international society and influenced its normative bases for centuries.

I do not wish to suggest, then, that there is anything inherently wrong with transcendental or transitional accounts of power. The goal here is *contrast,* in order to outline how counterpower is based on transgression. The brief landscape of examples covered so far that help categorize these perspectives suggests only that scholars have a decision to make when analyzing the when and where of power's instability, a decision determined by the scope of the research question one asks. For instance, if the inquiry rests on which organizational form(s) humans will value in the next few *centuries,* then a transcendental or even a transitional frame would make sense. As the aforementioned English School debates exemplify, if the investigation asks how it is that certain forms of international society wax and wane (Bull 1977: chap. 2; Watson 1992), if the world is transitioning toward a world society (Buzan 2004), whether the

---

14. This use of the word *hiding* also comes from Schroeder (1994: 430). In this case, it refers to the lack of joining one side or the other in a conflict.

15. Bull posits this as one of the three functions that the balance of power has historically played, to "prevent the system of states from being transformed by conquest into a universal empire" (1977: 111). Realist and English School perspectives on power balancing commonly assume that challengers to hegemonic power emerge for either societal or selfish (or both) reasons. Hegemonic power is challenged, but it is a choice, a consensus of the community, to do this, as it is not in the interests of international society for there to be an imbalance of power.

two orders will continue to exist alongside one another (Clark 2007), or even whether certain institutions (like the International Criminal Court) uphold (or upend) values relevant to either the international or the world society (Ralph 2005), then a transitional account of power, its meaning, and its effects or even purposes is quite appropriate.[16] Table 2 summarizes, in spectrum form, the temporal and spatial frames in transcendental, transitional, and transgressional analyses of power, with a few brief examples that will be conferred throughout the remainder of this chapter.

"A flash which loses itself": The Moment of the Transgression

Transgressional accounts rest on a philosophical ethos—a permanent critique of the Self and the manner in which we transgress the limits of what the Self is, a "historical ontology of ourselves" (Foucault 1984: 45). Transgressions are moments of infinite uncertainty, related to what Foucault titles the "limit," which is

> that narrow zone of a line where it displays the *flash* of its passage, but perhaps also its entire trajectory, even its origin; it is likely that trans-

TABLE 2. Temporality and Spatiality of Power

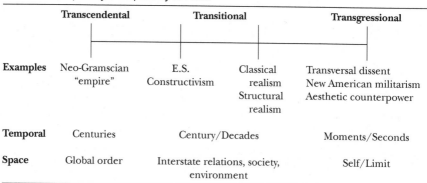

	Transcendental	Transitional		Transgressional
Examples	Neo-Gramscian "empire"	E.S. Constructivism	Classical realism Structural realism	Transversal dissent New American militarism Aesthetic counterpower
Temporal	Centuries	Century/Decades		Moments/Seconds
Space	Global order	Interstate relations, society, environment		Self/Limit

---

16. While I am hesitant to completely equate the categories in this chapter along the three "levels of analysis" as they have traditionally been understood in political science (see Singer 1961; Waltz 1959), I propose here that our research questions will determine how we "cut" into power's spatial and temporal frames. Indeed, we might view each of the following categories as examining processes that may intersect at the same time. The analysis in this chapter also engages how or if certain processes in one category can bring about the linkages found in another.

gression has its entire space in the line it crosses. The play of limits and transgression seems to be regulated by a simple obstinacy: transgression incessantly crosses and recrosses a line which closes up behind it in a wave of *extremely short duration* . . . But this relationship is considerably more complex: these elements are situated in an *uncertain context,* in certainties which are immediately upset so that thought is ineffectual as soon as it attempts to seize them. (Foucault 1977a: 33–34)

Foucault uses an apt metaphor, that the transgressional limit occurs "like a flash of lightning in the night," a "flash which loses itself in this space" (1977a: 35). This metaphor can be extended, of course, in that the shock of lightning comes in the broader context of a storm, heard from a distance in its thunder (which announces the shattered air of the light strike), but nevertheless a spontaneous limit that we can only know because of its *contrast* with the background. We ascertain the limit when it is experienced, and then we work back from that experience or rupture.

Why can we not think of these moments as any more than a flash or, over time, dynamic experimentations that to some degree defy extrapolation? In terms of the counterpower explored in this book, a societal power will swamp appearances of counterpower. It categorizes these microchallenges in order to control, mitigate, and exterminate their effects, as the brief discussion of Cindy Sheehan in chapter 1 illustrated (see also Delehanty and Steele 2009). Once a counterpower emerges over a lengthy amount of time relevant to its initial shock, it no longer contains its most potent force—its unpredictability.

As discussed earlier in this chapter, the regularity of transcendentalism and even transitional perspectives leads them to bracket large periods and spaces of stability, to take them for granted, to pattern them as the norm. Even if, for example, transitional accounts assume that the "state of nature is the state of war," they still propose balancing periods that prevent great power or even central wars (which are presumed to be primary in terms of analytical focus). In other words, if we analyze power's purpose and possession in more comprehensive spatial and temporal contexts, the notions of insecurity and disruption—of *struggle*—are de-emphasized. Transgressionalism instead positions struggle as perpetual. Thus, it accommodates an ethos of insecurity, as the Self is never firmly fixed. Temporal emphasis on the moment or days, rather than years, decades, or centuries, implies that power is always suspended and ready to be engaged, ruptured, or broken. It is never, in a sense, "overturned."

The contours of the "unit" only emerge during the transgression, which is so momentary that it defies the full formation of such a unit.

Transgressional analysis resists the temptation to scale up to the transcendental, "in the sense that it will not seek to identify the universal structures of all knowledge" and to instead turn away from "all projects radical and global" (Foucault 1984: 46–47). Instead, the transgressional seeks to identify the frontiers of our Selves as free but undefined beings. Or, as philosopher James Tully notes, a transgressional perspective is "not mean[ing] a total revolution or another view of the world but the *cautious experimental modifications of our specific form of subjectivity,* . . . [effects that] transgress the context by causing us to look at practices of [for instance] discipline and surveillance in the prison and in other practical systems in different ways and from different perspectives, *from the inside*" (1999: 98–99, emphasis added). This jettisons the notion of a pregiven or ever-constituted subjectivity, of a unified Self, and instead, in its place, sees a fragmented and always problematized subject (Simons 1995: 20). This is why Foucault uses the model of the artist to capture the ethos inherent in the transgressional form of analysis. Like the artist who creates, we cannot know the Self in advance; we can only observe it as it is in the act of being created (Osborne 1999: 48).[17]

"... in this space": Locating the Mark of the Transgression

What, then, of the space and struggle within the transgression? Recall that transitional accounts recognize a relational aspect to power, presuming that power exists in a community that imbues the meaning of which resources an actor can use to "carry out its own will."[18] Instead, in a transgressional ethos, the material and strategic acquisitions transfer from becoming an independent object used for the purposes of exerting influence on others to a property fused with the subject's identity—a correlative appendage of "the body" of the Self. Erich Fromm (1964: 162) titles this the difference between "the power over somebody" and "the

---

17. Yet such an unreliable state is the condition that allows for the experimentation and continuous creation of the Self. As I mention in the conclusion in more detail, a template of the Self is ultimately ambiguous and loose, and it is that ambiguity that the moment of counterpower engages and insecuritizes.

18. Patrick Jackson's development of relationalism analyzes phenomena as public spaces where rhetorical strategies are constrained and enabled by rhetorical commonplaces, although he concludes his 2003 study by stating, "The approach that I have outlined is in principle applicable to the analysis of *any* social action, particularly that kind of action that has as an effect the stabilization or transformation of actor boundaries" (249).

possession of power to do something." Such possessions that could be used to facilitate a transcendent order or even a transitional one become, at the same time, materials for the modification of the Self. As mentioned previously, the transgressional is more of an ethos or philosophy than, as Foucault states, "a theory, doctrine, [or] permanent body of knowledge that is accumulating" (1984: 50). Thus, this ethos of permanent critique lends itself to a creative approach to power—fashioned spontaneously and without recourse to societal codes or principles. Weber's definition that forms the basis for the relational aspect of transitional power—namely, the emphasis on the "opportunity to impose one's will within a social relationship"—can be contrasted with the notion that power implies the subject's ability to act creatively within its own series of active spaces.[19]

Transgressional perspectives both locate the "mechanism" for creativity within the Self of a centralized body of power and see the Self as a resource for the transgression. This may confuse some readers, for if we look at the examples from the previous empirical chapters, the *particular agents of counterpower* may seem, instead, to be the resource for the transgression. In discussing transnational terrorism, Osama bin-Laden's "Cynical parrhesiastic" force analyzed in chapter 3 is, in this respect, an example where the source of the transgression may appear to be him or his al-Qaeda network. Of course, there are certain material facts regarding al-Qaeda's generation of *physical* insecurity—such as its use of violence and the destruction that entails—for which it bears direct responsibility. However, the unfastening of U.S. power also emanates from the ability of al-Qaeda to use communiqués to challenge American *ontological* security. Thus, challenges to this sense of power came not just from al-Qaeda's attacks against U.S. targets throughout the 1990s but from bin Laden's 1996 fatwa that taunted the United States in its quick withdrawals from Beirut in 1983, Aden, Yemen in 1991, and Somalia in 1993. This was an explicit calling out of the masculinized component of the U.S. Self, a component archived by several theorists (Weldes 1999; Fierke 2007: 198; see also Cooke and Woollacott 1993).[20] It was thus a U.S. Self pre-

---

19. There is a surprising amount of overlap here between Foucault's notion of the transgression and the assumptions made, in slightly various ways, by Arendt and Parsons about the role of creativity in communal power, as noted by Berenskoetter (2007: 4).

20. Bin Laden stated in that communiqué, "You have been disgraced by Allah and you withdrew; the extent of your impotence and weaknesses became very clear. It was a pleasure for the 'heart' of every Muslim and a remedy to the 'chests' of believing nations to see you defeated in the three Islamic cities of Beirut, Aden, and Mogadishu" ("Bin Laden's Fatwa," http://www.pbs.org/newshour/terrorism/international/fatwa_1996.html).

occupied with "strength and will" that facilitated a transgressional context that al-Qaeda engaged with Cynic parrhesia. Therefore, as discussed in this book, counterpower used the aesthetic visions of the powerful target as *its basis* for its influence, rather than posing an imagined alternative. It is thus conservative in its ends.

It may be possible for actors seemingly secure in one space to be quite vulnerable in another *at the same time.* Take, for instance, Waltz's observation that in anarchy, one of the benefits of having power, or possessions, is that "the more powerful enjoy wider margins of safety." Powerful nation-states, Waltz explains, "can do the same dumb things over and over again . . . They can react slowly and wait to see whether the apparently threatening acts of others are truly so. They can be indifferent to most threats because only a few threats, if carried through, can damage them gravely" (1979: 194–95). Indeed, that may be the case in a transitional sense, where the goal is a balance of power. It may even make a nation-state primed for a transcendental overturning of international order.

But the possession of power, when fused, as noted earlier, with the Self, creates, in an inverse way, a greater potential for transgressions. Take, for example, Karin Fierke's exposition of Hurricane Katrina as a "*moment* when the vulnerability of this [U.S.] giant was exposed to the world." Fierke notes that "despite all its wealth, despite all its rhetoric about being the greatest nation on earth, about terrorists who resent the American way of life, despite its legacy of welcoming the poor and the oppressed, the vulnerability of America was exposed in the masses of people who were left stranded in New Orleans" (2007: 200). The inversion of power's vulnerability in the transgression is seen by simply replacing each of these "despites" and replacing them with "due tos," as in "*due to* all its rhetoric about being the greatest nation on earth."

Thus, would-be aspirants for transcendental ascendancy may, in fact, be more vulnerable to transgressional limits than they are to even transitional constraints. The United States, because it claims to be separate from international societal codes or laws—in fact, it even celebrates this perceived autarky on occasion, thus the references to America's "exceptional" historically geographic and idealistic position—finds itself privy to a self-creation because it perceives that it has transcended the "old world" of power politics. As Liska remarks, "having set out to reshape the cosmic order he [the aspirant] ends up *perverting* his communal order, since being ever more repressive internally will have become a condition of" such an attempt (1990: 190, emphasis added). Such a compromise of the aesthetic Self may or may not be intended—it may not even be justified or defended—but it can serve to soil the intrinsic and imagined beauty of that Self in the vain attempt to transfer such aesthetic to the system. In short, an agent may be transitionally secure while transgressionally anxious.

Such an aspirant power is not only constrained externally by an international society of actors that defend the status quo. It becomes constrained internally by the logic of the aesthetic Self. Transgressions that have been committed against this can be ignored or even obviated for a period of time—providing plenty of opportunities for scholars to uncover the organized forms of hypocrisy that permeate the discourse of great power states (see Krasner 1999; Weaver 2008; Bukovansky 2005). But if the aesthetic Self is an important component to the legitimacy and engine of such an aspirant's power, then a transgression also contains a disciplinary mechanism as well. The aspirant power has contradicted an internal code that, in some ways, is just as important as the external balancing that occurs in the society it finds itself at war against.

Yet if transgressional limits occur more frequently than, but still within, transitional processes, can the processes in the transgression impact those found in the subject's transitional capabilities? After all, we have already reviewed how transitional logics of power constrain actors (both nation-states and other transnational actors) in their move toward a transcendental order. While the same event may generate vulnerability in both a transitional and a transgressional sense, it is difficult to see many cases where repeated breaches of transgressional limits upwardly "cause," in a detrimental manner, an agent's ability to exist in a transitional order. This is so primarily because the limit—which is constructed through the process of the Self's ambiguity—is a limit valid for the Self, not the community. It is also because a hypothetical situation where a limit's break bears on transitional orders does not fully satisfy the requirements of "causality," at least in a classic sense.[21]

Even if we could not conclude that transgressions were important for

---

21. This is where an assumption of X (in this case, transgression) must exist independently of and temporally prior to Y (in this case, the transitional ability of a Self to get an Other to do what they otherwise would not). Neither is the case, as a transgression occurs *within* regimes of transitional power contestation. For this understanding of causality, see Wendt 1999: 79.

transitional or transcendental reasons, they would still be analytically interesting to investigate. Yet it is plausible that negative ruptures of the transgression can contradict publicized logics of the Self that the community holds to be important, or "legitimate"; and therefore transgressional limits may indeed *at the same time* impact the standing or authority of a Self in a community. Erich Fromm, furthermore, posits that the power to dominate is "mutually exclusive" from the power to be able to realize "potentialities on the basis of freedom and integrity of [the] self," a type of transgressional process, because an insecure Self will manifest toward the need to dominate. The response to a lack in the transgression, Fromm explains, is to seek out domination in the community (1941: 162). Of course, we can also sketch out scenarios where "possessions" that are useful for transitional protection in a community refine and modify a Self in a way that such a presentation obtains aesthetic legitimacy back at the level of the community, as I proposed in chapter 1.

Further, transgressional analysis configures a particular role for the intellectual. The transgressional uncovers the manner in which intellectual legitimation of transcendental orders leaves such orders inherently exposed, an argument I advanced in chapter 3 via the "liberal idealist" and with regard to academic-intellectual parrhesia. In other words, whereas intellectuals play a role in the *consolidation* of power in certain transcendental accounts, thus preventing counterhegemony from emerging, the relationship between intellectuals and power in a transgressional account makes power much *more vulnerable* to counterpower during transgressions. Much like the critical theory perspectives it uses as its inspiration, academic-intellectual parrhesia, for example, is a counterpower style that is perpetually opposed to power, exposing and uncovering its authoritative basis without proposing an alternative, expansive, and comprehensive "solution." Shaw and Russett's approaches elide the possibility that the intellectual, when finished tearing down the barricades, might become synthesized and then contaminated by power structures. Such co-optation not only would serve to eliminate resistance but would transform the idealistic purpose of the intellectual into the conservative purpose of the authoritarian. The emphasis on the "moment" in counterpower's transgressional basis means that it is, in contrast to the form of resistance found in transcendental accounts, *not* counterhegemonic. Because counterpower can occur at any time, its emergence defies point prediction, and the role of the intellectual in all of this is simply to better characterize those ruptures, rather than predict their occurrence.

The Role of the Other

As the examples in previous chapters suggested, counterpower as a form of transgressional analysis differs from the other two in terms of the role of the Other. For example, in Alexander Wendt's aforementioned thesis of a world state, the main driver is the struggle for recognition. We can only be recognized, according to Wendt, by an Other, and thus "subjectivity depends on inter-subjectivity" (2003: 511). This is the function of the world state—it serves to satisfy the universal will of all actors to be recognized. Similarly, in Shaw's thesis, an authoritarian Other stimulates a struggle for the global, democratic state.

Likewise, in transitional terms, the Other is there to help recognize and relate the meaning of power. It confers authority for actors "in recognition of the benefits an actor has provided for the community as a whole" (Lebow 2007: 124). The Other is a fellow *subject*, from and through which we also, as subjects, develop an understanding. Thus, an *intersubjectivity* of webs of meaning develop. Here, however, the ethos of creativity is not as present. When subjects become intersubjectively "locked in" to institutions and rules, they have sacrificed, in large part, this ability to self-create, in favor of a communal creative complex. When we are embedded in a community, we recognize that our actions impact those around us. Thus, while we might not constrain ourselves in those actions, we nevertheless justify, or "plausibly legitimate" those actions on *societal principles* (Wheeler 2000). In transgressional accounts, the Other has a different status. It only serves as an *object*, a canvas, through which we can assess the contours of the Self.

In contrast to a seminal work of post-structuralism (Campbell 1992), the Other in transgressional analysis is not the only concealed subject. What executes the power of the Self so forcefully is its own operation in darkness and ambiguity. While surely influential in its subjectivity, the examples from chapters 2–4 of the current book demonstrated how the Other (tsunami victims, Osama bin Laden, Fallujah insurgents, or My Lai victims) was still not as vital to the Self's identity as it is depicted by Campbell. What is important is that the presentation of the Self, facilitated by power's need for movement and action, escapes its own boundaries—the Self could appear anywhere and everywhere because of its search for self-meaning. While Campbell's thesis stands largely correct, the reason for this insecurity results from power's subjectivity as an aesthetic creation. Such work still implies a perpetual "movement" of the Self, but a movement formed within and through a previous Self, rather than solely against the risky internal or dangerous external Other.

We thus might consider a possibility that in certain contexts, recognition by the Other is not necessary for the Self.[22] If there was an Other in the formation of an integrated Europe, for instance, Ole Wæver suggests that it "can be distinguished in time, rather than space: not Russia, but Europe's past must be negated. This is reflected in the rhetoric that Europe has to be integrated, otherwise we will fall back into the power balancing and rivalry of the former Europe" (1996: 122 n. 14). Without an Other, the Self must be constituted within the "open but heterogeneous polity" of Europe itself.

The melding of power to an ambiguous (or nonexistent) Self is particularly delicate when a previous incarnation of the Self is not fully formed. For a period of time during the twentieth century, it was fashionable for some realists, such as Reinhold Niebuhr, to, in such a vein, view the United States in terms of its "maturity" level as a great power.[23] During the Spanish-American War, Niebuhr notes that American hypocrisy was "probably excessive, because a youthful and politically immature nation tried to harmonize the anti-imperialistic innocency of its childhood with the imperial impulses of its awkward youth" (1932: 88). Such transgressional vulnerability becomes stark when assessing the "maturity" of U.S. power. Materially, that the United States became a "great power" only a century after its birth made it less reliant on "soft power" measures such as diplomacy and moral authority and, instead, much more dependent on military force as a first resort. Because of its childlike emotional status and its assumption that U.S. ideals were universal ideals, U.S. power may be more susceptible than more "mature" powers to what I (Steele 2007c; 2008a: chap. 3) and Andrew Linklater (2007) have termed "shaming" processes. Without an Other against which to formulate most, if not all, logics of the Self, an immature Self is inherently fragile. Indeed, Niebuhr notes that America "was just beginning to feel and to test its strength and was both proud and ashamed of what it

22. In fact, however, Wendt does seem to assume the Self's work on the Self, eventually, because once a world state emerges, "short of an extra-terrestrial Other," there is no further recognition possible. For this, one of the constitutive "drivers" Wendt uses is a past Self: "A world state could compensate for the absence of spatial differentiation through a temporal differentiation between its present and its past" (2003: 527). Yet as Wendt also admits, while the reverse is true, the past can not recognize the present, and thus his main driver is no longer present at this stage.

23. Morgenthau agreed with this general proposition but differed when it was applied to the United States, arguing, in 1950, "The United States offers the singular spectacle of a commonwealth whose political wisdom did not grow slowly through the accumulation and articulation of experiences. Quite to the contrary, the full flowering of its political wisdom was *coeval* with its birth as an independent nation" (833, emphasis added).

felt" (1932: 88). Foucault's views on the care of the Self, discussed in chapter 1, also portrayed the young as incapable of exercising the full ascetic techniques necessary for this care, how the "adolescent's entry into life, his transition to the phase we ourselves call 'adult,' poses problems" as a "dangerous passage" (Foucault 2005: 87).[24]

Toward a Transgressional Account of (Counter)Power

A few analytical comments are in order here as we conclude this general proposal for the transgression and before moving, in the final section, to some examples. First, how can we know a transgression when it occurs? The use of the word *goals* by Wolfers to refer to possessions and the milieu implies that we can ascertain or at least model the *intentions* of global actors (such as agents for the nation-state). In the transgression, however, intentions are less relevant for analysis because the transgression can occur within transcendental and transitional orders: we can find actors having particular intentions for power but nonetheless become insecuritized for logics that are largely irrelevant to the particular order of goals for which they prioritize power.

Transgressional analysis concerns the *performance* and *appearance* of the Self and the *effects* of counterpower, rather than intentions. It matters little what the sense of power was acquired for. It matters little as well what the initial extension or execution of a performance was intended for "in the beginning." What matters instead is how that performance creates a particular mode of being. This mode of being can then be disturbed, and it is the disturbance—"something that we can empirically grasp" (Jackson 2006: 24)—that we seek to explicate. Yet the investigative model for this is, as expected, more art than science. Transgressional analysts must *sense* a disturbance. We hear it or see it; we, in a sense, feel not the Self but the lack of that which once was. What once was is still there, of course, as it has been not upended or transmogrified but disturbed by the nothingness being revealed.

Because the transgression is not linked to an external, universal struc-

---

24. Obviously there are different assumptions embedded in Niebuhr's understanding of an immature U.S. Self and Foucault's indexing of a Self in constant creation. One of the insights gained from incorporating generational analysis, as I do in the conclusion, is that it accommodates, to some degree, both the skepticism of entrenched regimes of power who view younger "upstarts" as being immature and, at the same time, how generational shifts lead to an ability for self-creation at the communal level, where the vanguard generation can engage the idea of a communal Self as new and where each vanguard engages that Self as if it is a clean canvas or frontier.

ture, it does not represent an independent reality. We must resist, in Roland Bleiker's words, "the desire to ground our existence in an external source—if not in God, then in something else that could take over his stabilizing position" (2000: 106).[25] A picture or image of torture's effects, for example, is different than the effects of torture on the body. The latter is a medical-biological-anatomical reality ascertainable through an autopsy or medical examination. The former is a political-philosophical disturbance of an aesthetic strata that the analyst must view and then explicate in order for it to come into "being."[26] It is what Patrick Jackson terms a "practical transaction that the researcher—and the research community—engages in" (2008: 142). This means, again, that we do not have to necessarily unlock intentions or motives in order to ascertain a "driver" or "predictor" of power's aesthetic subjectivity. I may have a "hunch" as to what motive power is pursuing here, but I can only *intuit* what it values in terms of aesthetics by observing how power reacts when those values are compromised. After observing the way in which U.S. power handles the transgressional "limit" in the examples I analyzed in chapters 2–4, it still proves difficult to establish the "true intention" of those actions that rectified the rupture, as it proves difficult to establish any intentions (Jackson 2006: 24). The most we can determine regarding intention in those case illustrations is that it was overdetermined. The U.S. reactions observed during these ruptures served objectives that were about attending to repairing a damaged U.S. Self. But what we can determine is how power reacts to a counterpower rupture. The "intentions" of power or even a counterpower usher are largely immaterial.

A final point to make is that the model of this aesthetic Self applies not only to the United States or even to broader "great powers" or only to nation-states. We must remember that Foucault drew this model of transgressional technique from the individual and then moved toward the metaphor of the painter because of the creative process and the

25. While it is beyond the purview of my current project, I think there are at least some stylistic similarities between Bleiker's Focauldian rejection of the transcendental and Patrick Jackson's Weberian-influenced critique of "building claims about the transcendental normative validity of a policy into the causal mechanism used to explain how that policy came about" (2006: 19), a critique Jackson (20) levies against the work conducted in the past fifteen years by constructivists such as Audie Klotz, Martha Finnemore, and Neta Crawford, as well as the work by liberal institutionalists such as G. John Ikenberry.

26. "Comporting oneself towards a river as equipment for recreation discloses a different river than does a comportment that regards the river as a source of raw energy to turn a turbine" (Jackson 2008: 142).

"ethics of existence," which could be applied to any being in which aesthetic integrity must be constantly reinforced. Such an ethos obtains in a multitude of organizational bodies—and I think this makes intuitive sense when we consider the amount of time, energy, and money that corporations, for instance, spend crafting an aesthetic image or images to promote their products. But in case it does not appear intuitive, we have at least one study, by Alison Hearn (2008), that comes to us from the field of information and media studies and asserts a "branded self" with reference to the body or mode of being, which is used by reality television programs. Popular U.S. television shows such as *The Apprentice* and *American Idol,* for instance, are "commodity signs" that point to the branded Self and are thus self-referential firmly because there is no stable basis for their existence (Foucault 1977b).[27] As Hearn also notes, social "networking" sites such as MySpace and Facebook also stimulate uses to brand the Self, and although they are *social,* they depend on the author to constantly update, create, and revise their online Self "profile" (2008: 210–13).

Transgressionalism and International Relations

In order to better illuminate just what transgressional accounts of power look like, I here present two examples that illustrate some transgressional themes. The first is Andrew Bacevich's (2005) thesis on "the new American militarism," and the second lays out how all three categories of power contestation reviewed in this chapter might investigate an incident occurring in the recent Iraq War.

As detached and esoteric as the transgressional may seem, it is in fact a view that is illustrated, first, through Andrew Bacevich's (2005) conventional, self-titled "Conservative" and "Catholic" account of U.S. power and identity. In this study, the supposedly strategic resources of power, or the "possession goals" that nation-states acquire, bonded into the fabric of the national Self and became that which could be "transgressed." Bacevich's thesis is that the United States has embraced military power as a first resort and that such enamoring occurred "as a reac-

---

27. These are not, obviously, fashionings that may deny or ignore the societal links with the Other in their various spheres. Sometimes they are in competition with Others, and thus their operational existence depends on relative and strategic goals for their brands. The branding and then commodification of the "state" is the subject of Jeremy Youde's investigation into the "selling" of South Africa within the confines of an "African Renaissance" (Youde 2005).

tion to the 1960s and especially Vietnam [and] evolved over a period of decades" (6).[28] Bacevich defined this militarism as "the misleading and dangerous conceptions of war, soldiers and military institutions that have come to pervade the American consciousness" (ix), and this contains at least four components—each of which speak to how deeply militarism has been incorporated into the U.S. Self. First, there is the "scope, cost and configuration of America's present-day military establishment" (Bacevich 2005: 15). But this first component is more profound, as it is more than just the relative (to other periods of U.S. history) material costs of this new American militarism. The military establishment has gone, as Alexander Wendt (1999: 250) might observe, from the second to the third degree of norm *internalization,* where such a norm actually constitutes who an actor *is.* As Bacevich observes, these costs "elicit little comment, either from political leaders or the press. *It is simply taken for granted*" (2005: 17, emphasis added). Two further components of the new American militarism include the "normalization of war" (ibid., 18) and "an appreciable boost in the status of military institutions and soldiers themselves" (ibid., 23).[29] The former is characterized by a comfortable sense of war by the public and politicians, who read daily about soldiers deployed in several theaters of battle. The latter has essentially elevated the soldier to a status above the citizen. "Paying homage to those in uniform has become obligatory," writes Bacevich, "and the one unforgivable sin is to be found guilty of failing to 'support the troops'" (24).

But these three together are not enough to constitute the new American militarism, for it tellingly also includes what Bacevich calls "a new aesthetic of war." Whereas twentieth-century aesthetics of war portrayed war as "barbarism, brutality, *ugliness*" and portrayed the military as "an inherently degrading experience and military institutions by their very nature repressive and inhumane," the new aesthetics saw war as a spectacle where "appearances could be more important than reality" (Bacevich

---

28. Jutta Weldes (1999: 50–51) similarly avers how the emphasis on "strength" in the U.S. Self had been cultivated since at least the Vietnam War and perhaps even before.

29. Francois Debrix helps provide further explication on what has brought about the last three of these components in his book *Tabloid Terror.* Debrix's observations about the U.S. Army's shift in recruiting tactics as both a consequence and further reinforcing practice for this "military theme in American culture" bears notice here: "Increasingly these days, the Army presents itself to young Americans not just as a war-fighting and soldier-making industry, but also as an educational and professional training institution, one which, unlike most American universities and higher learning institutes, will provide its members with immediate real-life experience" (2007: 92).

2005: 20, emphasis added).[30] War now was "palatable . . . grand pageant, performance art" (22).

Two observations can be made about the import of Bacevich's thesis for the transgressional approach to power. First, the new American militarism is deeply expanded. It does not include just one individual or institution but pervades the fabric of U.S. society. Thus, power here interweaves into the U.S. Self, yet this Self cannot obtain a sense of objectivity (as a work of art being viewed) until it is seen *in action,* and it is only checked when it transgresses "the limit" (a view of aesthetics as action explicated in my discussion of vitalism in chapter 1). Moreover, because it has been interwoven more deeply than only through one individual person, party, or other sector of U.S. society, it cannot be expunged as easily. To put it in terms of the aesthetic "strata" advanced earlier in the book, such militarism has infested the rhythms, psychology, and imagination of U.S. citizens, or the aesthetic "strata" of U.S. power. This possibility bears critically on the faith that democratic peace theorists have in the notion that the "multiple power centres of democracies make abrupt and untoward state actions more difficult—sharp change in policy requires more actors and institutions to sign up to it than in non-democracies" (Ikenberry [2000] 2008: 37), for if power has been entwined to U.S. society, who will get in the way of potential "sharp changes"?[31]

A second observation regarding the new American militarism thesis—one that bears even more on the transgression—is that Bacevich sees the extent of its development as due not to a rational external threat (which would presumably promote a "security dilemma") or even to a

---

30. The editors of *Millennium*'s special issue titled "The Sublime" (2006) note about World War I that "it is this war that is most associated with poetry, for it is only through poetry that the reality of conditions at the front could be translated to a public entirely unfamiliar with the scale and destructive power of this new kind of warfare" (ii–iii).

31. One of my favorite hobbies is to view the analysis that has come out in only the past several years, since the 2006 midterm elections in the United States switched control of Congress to the Democratic Party, and examine the prognostications regarding what such a switch would presumably have for the Iraq War policy. Those who have faith in the democratic process, like Michael J. Williams (2007: 274), read the events of April 2007, when "the Democratic-controlled Congress was in the process of putting legislation to the President that mandates withdrawal of US forces over the course of two years," as if such withdrawal would really occur. Instead, in that same year, the U.S. increased forces through a troop "surge" in Iraq by a total of at least twenty-four thousand combat forces. Apparently, the heroic "loyal but independent political opposition" assumed in democratic peace accounts, which is supposed to "limit the ability of a democratic government to wage a capricious, ill-advised war" (Russett and Oneal 2001: 248), also does a poor job at stopping such a war after it has already been commenced. Perhaps there is more at work here than just domestic politics.

shift in the global distribution of power but, rather, to a previous unfavorable manifestation of the U.S. Self as it existed through the Vietnam War. Bacevich notes,

> Militarism qualifies as *our very own work*, a by-product of our insistence on seeing ourselves as a people set apart, *unconstrained by limits or by history*. More specifically, in this case, militarism has grown out of the Vietnam War . . . Vietnam was a defining event, the Great Contradiction that demolished existing myths about America's claim to be a uniquely benign great power and fueled suspicions that other myths might also be false. (2005: 34, emphasis added)

Vietnam was an object through which the U.S. recognized its own transgression. To be clear, it was not the "whole" experience of Vietnam that was the limit for the U.S. Self. Rather, the process of the "flashes of lightning" (to paraphrase Foucault from before) encircling the U.S. Self as it exists within Vietnam established this limit. However, as Bacevich also notes (ibid.), Vietnam created two forces of reaction, and thus two logics competed for the space of a reformulated U.S. Self. In chapter 4, I analyzed this division through the "flashes" of two Vietnam experiences—the My Lai massacre and the Fall of Saigon. There is, to borrow Ole Wæver's notion, a "*self*-negating, *self*-transforming argument" that obtained regarding the hole of Vietnam (1996: 122).

Digging into Fallujah: Where Transgressional Counterpower Begins

For a final illustration on when and where the analysis in the previous chapter was located, let me use the Fallujah example discussed in chapter 4, an event that can be analyzed using any one of the layers discussed in this chapter. Using a transcendentalist lens, the event could be considered part of a structural challenge the world faces in its progression toward a global federation of democracies, with the latter being a stopover toward a world or global state (Wendt 2003; Shaw 2000, 2001). The location of this struggle is important as well, as certain democratic peace scholars see areas of the globe as rough patches for the proliferation of democratic "networks" (Maoz 2001). If such "hard case" regions as the Middle East could be democratized, this would serve as an agent or springboard toward global democratic transformation, a key peg in the disciplinary logic of systemic democratic peace theory (see Steele

2007a). A neo-Gramscian might point to the occupation of the contractors—not soldiers but, rather, members of a private security firm—as evidence of the "wave" of privatization that began in the 1990s as a transferring of "risk" from the nation-state (in this case, the United States) to a private firm (Blackwater).[32] To understand this condition of flux, one must reconstruct how the global forces of production (including ideational forces) manifest the kinds of policies that allow (even necessitate) private security contractors to be used in a theater of modern war. Could this moment of carnage precipitate or be the initial result of a shock that delegitimizes this emerging form of neoliberal economic world order? Finally, perhaps Fallujah is yet another flash point of a clash of civilizations—an example of the fight between the West and the rest, or, as the head of the Coalition Provisional Authority said at the time, an example of "despicable, inexcusable and barbaric acts" that violated "the foundations of a civilization" (Cornwell 2004). Thus, Fallujah was a fault line between an emerging Islamo-fascism and a U.S.-led vanguard of force to demonstrate to "thugs" that their celebrations would not go unpunished (Bay 2004).[33]

On a transitional level, the celebration of Iraqis may demonstrate for realists like Mearsheimer (2005) how a neoconservative policy of preemption would promote, even hasten, balancing rather than "bandwagoning." These realists, many of which were in the "vanguard of the opposition" to the Iraq War (Walt 2006), would examine Fallujah as an almost anticlimactic example of the depletion of U.S. power and an indicator of U.S. relative power decline. Neoconservatives would (and did) argue that the precise moment of Iraqis celebrating necessitated a reaction to reassert strength in the face of adversity. Only such a reaction can engender emulation, if the United States is to act as the world's "chieftain"; otherwise, it would appear to be weak and thus not a worthy ally or model of democratic development (Kaplan 2002: 132). At a more tactical level, the Fallujah uprising can be considered just one part of several "Iraq wars"—occurring alongside the battles that became more acute that spring of 2004 in the form of the Shia uprisings and in the fight against an increased transnational terrorist network affiliated with al-

---

32. Private military contractors (PMCs) are of course becoming a hot topic of analysis for IR scholars (see Singer 2003; Avant 2005; Abrahamsen and Williams 2007), but to my knowledge, an as-yet-unexplored but fertile terrain exists for scholars who would wish to explicate PMCs from a neo-Gramscian perspective.

33. See http://www.strategypage.com/on_point/200446.aspx.

Qaeda and led by Abu Musa'b al Zarqawi. For English School pluralists, perhaps Fallujah (in a degree or two of separation) also represents the disorder that comes about when the norm of sovereignty is compromised or, even worse, when one of the great power "custodians" (see Bull 1977: 19) instead became a rogue member of international society that compromised the sovereign norm through preventive war.

But we may also view the prevalence with which the "ghosts" of Mogadishu were invoked by U.S. policymakers and investigate in which location this particular vulnerability arises. If we can ascertain such a location, we might also ascertain its particular threat destination. But to do this, we would need to engage the transgressional level. The specter (in an almost literal sense) is not of an Iraqi insurgency but of a "wound" in a past U.S. Self, where "the pain is attributed not to the act of destruction, but to its memory" (Zehfuss 2007: 124–25). A U.S. Self whose aesthetic basis has been compromised helps explicate the surprisingly larger concern that U.S. policymakers, commentarians, and military leaders invoked of American ontologies past rather than an Iraqi insurgency of the (at that time) present.

A final thought that the Fallujah disorder and the subsequent U.S. decision to reactively attack the city shortly thereafter illustrate, as does Bacevich's analysis on the materiality of the Self, is how they provide a possible outcome of the Self's transgressional work that Foucault did not anticipate when he linked such "games of power" to freedom. If the transgression occurs in the context of constant struggle, then the freedom is really a dizzying context of imprisonment. Internalized power is transgressed through a constant working back on the Self and, when challenged ontologically and aesthetically, forces the Self to re-act against this challenge. It is this vulnerability that previous chapters have excavated and explored. Gone largely unearthed, however, is whether the vulnerability, insecurity, unpredictability, and fragility of power is a *normatively beneficial development,* one that should be championed, or whether the permanent upending of the status quo will lead to a disorderly world where rebellion becomes the rule, rather than the exception, and where humanity's paranoia follows logically in its wake. These are the issues I confront in the conclusion of this book.

Conclusion

In a famous essay written during the Battle of Britain in 1941, George Orwell wrote what I find to be a very fitting quote with which to commence the final chapter of a book on aesthetics and power. Orwell's essay begins by discussing the cultural differences that existed, at that time, between the English, on the one hand, and the Germans or Italians, on the other. One difference he mentions is that the English "are not gifted artistically . . . [and that] they are not intellectual. They despise abstract thought." While this lack of artistic talent, in turn, breeds a resistance to aesthetics, Orwell observes that the English possess a "love of flowers." Rather than a love of aesthetics, the love of flowers is instead a symptom of the English's love of liberty, with the "*privateness* of English life" being no different than the English's preference for a "nice cup of tea."

Yet Orwell writes some of this tongue in cheek, for while the English may not have been capable of producing the beauty found in Italian and German art, they must have appreciated beauty enough to know what it was not.

> The goose-step, for instance, is one of the most horrible sights in the world, far more terrifying than a dive-bomber. It is simply an affirmation of *naked power;* contained in it, quite consciously and intentionally, is the vision of a boot crashing down on a face. Its ugliness is part of its essence, for what it is saying is "Yes, I *am* ugly, and you daren't laugh at me," like the bully who makes faces at his victim. (Orwell 1941, emphasis added)

What I find interesting in this particular passage is how it represents, quite nicely, some of the analytical themes that were discussed in this book. Power itself has an ontology, even an aesthetic manifestation. It *must* have this aesthetic element, for when it is "naked," it is ugly. It cannot have the effects or influences it does without such aesthetics. Eventually, such naked power facilitates its own resistance. Orwell's ode to "ugliness," as stylistically as he presents it, is by no means the only reference of its kind—for whether it be the allusions to "sickness" and "stains" issued by U.S. politicians regarding Abu Ghraib, discussed in chapter 4, or the "foul stain of our species" that Kant mentioned regarding radical evil in his work on aesthetics (Lilla 1998: 12), the reaction that follows beauty's defilement exhibits an impressive shock that needed to be explicated. Hopefully this book has gone some distance toward doing that.

By utilizing insights from several theorists and philosophers, this book has suggested that the subjectivity of power seeks out an aesthetic sense of self-integrity through constant creation and innovation. Put another way, the Nietzschean "will to power" comes coupled with a "will to beauty." This need for creation and modification of the Self and the idealization of such aesthetic integrity make such power inherently vulnerable to particular practices of counterpower such as reflexive and flattery discourse, parrhesia, and self-interrogative imaging. These manipulations, occurring in moments or days, cannot be predicted or preempted, and they give no warning, because it is the Self of power that facilitates, in part, their occurrence. By revealing this Self in an aesthetically unfavorable light, counterpower entails a shock that forces power to re-act to reestablish some form of aesthetic integrity. The illustrations used throughout this book, especially those of chapters 2–4, sought to demonstrate how this activity obtains through a variety of U.S. foreign policy practices.

The thesis that power seeks out an aesthetic totality that can never be realized and thus can be manipulated was largely developed as an analytical position. While I disclosed in the introduction to this book some of the contemporary events that have fueled my interest in this subject, gone mainly undisclosed is my own normative view on whether such insecurity is a positive condition in global politics. If my thesis about counterpower, aesthetics, and insecurity is correct, how should we go about resolving or at least constraining some of the more ethically knotty avenues down which counterpower takes us? We may not, for instance, easily embrace the ethos of counterpower if it manifests itself in the form of constant insurrection, where an authority that could provide order to a community is instead constantly delegitimized and therefore reacts to these challenges in difficult ways. Should we not, for instance, find it problematic that those who can wield organized violence do so because of some seemingly narcissistic drive of aesthetics? Or, regarding a case illustration presented in chapter 4, should we not be disturbed that U.S. soldiers and a still indeterminate amount of civilians in the Fallujah area had to die in a massive assault of that city in April 2004 because the United States did not want to *appear* as if it was as weak as it was portrayed in Mogadishu, Somalia, in 1993?

Alas, many of these ethical dilemmas cannot be resolved in this conclusion, but we begin to access the normative predicaments by first acknowledging the analytical reality. Aesthetics are important to power and its operation: they are important to elites who act on behalf of their communities, and they are important for scholars who are seeking out orderly paradigms of knowledge production. They are important for a member of any political community that idealizes a community Self. Far from solely being only an insidious development, the "shock" or "disturbance" of a counterpower can also be productive. Jan Egeland's stingy comment (in part) helped formulate a response by the United States that entailed at least the move toward delivering more aid to tsunami victims. The images of My Lai forced the U.S. military to reform many of its practices in the years that followed. The images of Abu Ghraib, far more than any well-intentioned but nevertheless mundane international conventions or national laws, helped shock the United States and caused a national debate to unfold over the status of detainees held in the War on Terror, a debate, admittedly, that the United States continues to face today.

Further, even though the quandaries that result from this reality cannot be resolved in this conclusion, it will nevertheless begin by suggesting that counterpower events can coexist with formations or principles that seek to find productive opportunities within a field of power relations. It first reveals the inherent difficulties with "one-size-fits-all" comprehensive solutions to global governance, as evidenced by Robert Keohane's 2000 presidential address to the American Political Science Association. It next posits one principle (restraint) that confronts the quandary about perpetual insecurity in a way that acknowledges the limits of human relations. It then designates the future study of *acupunctural formations*. Such formations, while containing several of the constitutive traits of counterpower (spontaneity and spatial minutia), also productively heal, if only on a localized level. This conclusion presents some possible philosophical resources that might explicate the occurrence of these formations.

"Clearly the stakes are high: no less than peace, prosperity and freedom"

How does one stem the velocity of power's aesthetic reaction? What is the answer to this problem of insecurity—of power and the authority of nation-states and even forms of global governance being challenged? One proposal would be to "build a better mousetrap"—to conceptualize and construct ways in which the aesthetic drive of the Self can be contained and channeled toward progressive or productive avenues for the global polity. For those scholars who study global politics, a sphere where violence is still practiced and where the threat of its use is still common, the tendency to seek out an urgent, comprehensive, and perpetual solution to such violence is admittedly justifiable.

For Robert Keohane, an iconic scholar who has served as president of both the International Studies Association and the American Political Science Association, the task is up to political scientists, who face a "challenge . . . resembl[ing] that of the founders of the United States: how to design institutions for a polity of unprecedented size and diversity . . . Political science as a profession should accept the challenge of discovering how well-structured institutions could enable the *world* to have 'a new birth of freedom'" (2001: 1, emphasis added). Keohane's ambitious proposal provides one avenue we might take in order to confront the problems of insecure power: develop institutional structures that can protect people from "self-serving elites and from their [own] worst impulses" (2). Invoking Habermas, Keohane proposes that we should seek to move toward a world where communication and persuasion "change others' minds on the basis of reason, not coercion, manipulation, or material sanctions" (2). In short, Keohane's argument boils down to constructing assessable institutions so that progress is not only possible but likely in the form, if not the function, of the U.S. experience.

I am skeptical to travel down this avenue, for several reasons. But let me first justify why I chose it as the basis for contrast here. I picked Professor Keohane because he is one of the dominant IR theorists of the past three decades, publishing two of the seminal works that redefined the field of IR theory—*Power and Interdependence* (1977, with Joseph Nye) and *After Hegemony* (1984)—along with a co-authored account on qualitative research methods, *Designing Social Inquiry* (1994, with Gary King and Sidney Verba), at least a portion of which nearly every U.S. graduate student studying international relations over the past decade has been assigned. On a professional level, if we are to find an "ideal" of professional accomplishment that budding scholars of international relations should seek to strive for in their careers, Robert Keohane can surely serve as this model.

Yet this address perfectly exemplifies, in my view, the necessity assumed by many current IR scholars for constructing some "plan" that can fix global problems once and for all. The idea that we should seek solutions on a *global* scale is nothing new and goes back to at least Grotius (with the advent of modern international law) and Immanuel Kant, through the twentieth-century plans for a world federal body, to the many current works that discuss plans for a global federation or even polity in various forms. Keohane's call represents for me the "urge" that many scholars have for *comprehensive* plans to provide the solutions or at least ameliorative frameworks for any or all of the problems humans face on a global level.

Armed with some of the results that stem from the counterpower account I have developed in the previous chapters, I would like to issue a series of critical questions regarding Keohane's "task," before moving on, in the last part of this conclusion and book, to some alternative avenues we might go down. To begin, I wonder why constructing these institutions should be the responsibility of political scientists or any of us who study the world. Is this truly our calling? If we do it, must we not take some of the blame for problematic institutional outcomes? In addition to holding elites accountable, as accountability is one of Keohane's "three procedural criteria for an acceptable global governance system" (2001: 3), should we not also ask for disclosures and even mea culpas from scholars when such avenues prove to be poorly traveled?

To clarify, I have specifically in mind here those scholars who advocate theses that have been used by international actors to justify abhorrently unproductive ventures, such as the Bush administration's use of democratic peace theory to justify Operation Iraqi Freedom (see Steele 2007a).[1] If global forms of governance, even those that embody the ideals used to measure "progress," instead lead to the "exploitation and even oppression" that Keohane is also concerned with (2001: 1), who, then, do we hold accountable besides the individuals who run these institutions? Must we not cast a wide net? For if we as political scientists are to go about consulting, even commanding, the global community on the

---

1. The ultimate evisceration on this comes to us from Friedrich Kratochwil's recent declaration that "the universalist dreams of a revolutionary change that catapults man from the realm of necessity into that of freedom are phantasmagorical eschatological speculations" (Kratochwil 2007: 41).

direction it should seek in terms of its own development, must we not bear some of the blame when things go awry in this plan? When we seek out order through institutions, we must also assume that such order will come part and parcel with the need by power to *control* unpredictability and creativity—concepts at the heart of counterpower. This is an epistemological challenge as much as it is a normative one—for if the scholar impacts the world she or he studies by constructing constitutive concepts that *re*-present that world, then she or he not only is implicated with its maneuverings but helps bring about its operations, both good and bad (Williams 2004: 23).

A second set of questions I ask is premised by a humble view of human abilities. How are we to trust that these institutions will be executed faithfully? Keohane closes his address by admitting that "if we bungle the job, the results could be disastrous. Either oppression or ineptitude would likely lead to conflict and a renewed fragmentation of global politics" (2001: 12). This is the characteristic my generation has faced, for we have witnessed ineptitude from national and international political and economic leaders on an almost unthinkable scale. It could be the United Nations' handling of the 1994 Rwandan genocide or the corruption that surrounded its Oil-for-Food Programme in the late 1990s and the early part of the current decade. It is located through the "hypocrisy" one finds in global institutions such as the World Bank (Weaver 2008) or in the Bush administration's response to Hurricane Katrina, which "exposed the relationship between power and vulnerability" (Fierke 2007: 199). It is evidenced by an invasion of Iraq whose stated intent was a concern over weapons of mass destruction but that left weapons depots in that country unguarded and unexploded because they might have WMDs in them (Ricks 2006: 145–46). The point here is that it is not only "the worst people" who thrive "behind grand plans" (Keohane 2001: 7). Good people may as well. Well-meaning incompetence breeds misery all the same.

There are some mechanisms of accountability, as Keohane suggests, that could stem this incompetence. But Keohane's primary one, democratic accountability, has, in recent years, taken what we can only consider to be a bit of a drubbing. How accountable has the world's first "modern" democracy held its leaders for their *stated* use of torture against individuals detained in the war against terrorism? Nondemocratic forms of accountability that Keohane lists—such as the global market (2001: 9)—have also lost their luster in the past several years, with U.S. financial institutions receiving nearly one trillion dollars in "bailouts" from the U.S. government even though, for several years, the former promoted a quasi-Ponzi practice of subprime mortgage lending that could never be sustained. Who, again, has been held responsible by this market? I would suggest, in telegraphing my proposals for acupunctural formations, that accountability mechanisms are most effective when they are personal and thus local. The values we should uphold are not only good people or good institutions; they are also competence and pragmatism. Those values are difficult to maintain with grand, one-size-fits-all visions of global governance, even if they contain mechanisms for democratic accountability. Or, as Morgenthau stated, in an essay aptly but skeptically titled "The Escape from Power" and in terms that invoked the staggered and uneven flow of generational change, "political problems . . . cannot be solved by the invention of a mechanical formula which will allow mankind to forget about them and turn its attention toward a not-yet-solved political problem. Being projections of human nature into society, they cannot be solved at all. They can only be *restated, manipulated, and transformed,* and each epoch has to come to terms with them anew" (Morgenthau [1947] 1962: 313).

Facing these limitations, an alternative practice to promote here is not accountability but transparency and an angry form of skepticism. We should consider in this vein one of the results that comes from the aesthetics of insecurity via the trope of power's "dark" operation, noted especially in chapter 4. Foucault's (1984: 138) conclusion (also mentioned in chapter 4) that bears especially on the imaginative stratum of aesthetic power is that our love of an object is most intense in darkness. Power takes advantage of this love because it is not seen—it is only when the lights are turned on that we become "wounded by the unloveliness of the images." The eros of such love is lost. We cannot forget how ugly power, well-intentioned or not, is when it operates in the dark. Our vigilance requires not our holding such power accountable through democratic or market mechanisms but an angry skepticism sluiced through perpetual critique and exposure of the aesthetic subject of power. Once exposed, power may work back on itself, forcing it to reform its operations if it knows it is being "seen."[2]

---

2. There is yet another intersection to note between Morgenthau and Foucault, for on this point, Morgenthau stated in the prologue of his *Truth and Power* volume, "The experience of the 1960s has dispelled the illusion that truth can show power the way in direct confrontation. But historical experience reassures us that truth can indeed make people 'see a lot of things in a new light.' And when people see things in a new light, they might act in a new way" (1970: 9).

A third set of questions centers around power relations. Does Keohane's plan ignore them or, even more audaciously, seek to, in Morgenthau's term, "escape" from them altogether? The nod to Habermasian forms of persuasion does not sound like a promising recognition of power's perpetual operation. The aforementioned castigation of processes of "manipulation and coercion" also, it appears, provides no haven for power in a future global polity. It is obvious from what I wrote in chapter 2 regarding such models that I consider the process of communicative persuasion to be highly problematic. I also think we should not view such a process of persuasion as divorced from power-based notions of "force" or "rhetorical coercion," as the works of Bially Mattern (2005, 2007) and Krebs and Jackson (2007), among others, have helpfully articulated.

To the extent that it is bound up with legitimacy, which is another important ingredient for Keohane (2001: 3), I wonder what we are to conclude from the fact that the Bush administration could legitimate the torture of terrorist suspects to a democratic U.S. polity by *persuading* them that it was necessary to foil future attacks? Legitimation also occurred in the international community, where the United States supposedly convinced thirty nation-states that Saddam Hussein's Iraq, a country monitored by two no-fly zones for twelve years, still posed a threat so great that a preventive war was necessary by 2003. The legitimating principle that is used to persuade a community is dependent on the logical faculties of the community in which it is legitimated.

A fourth set of concerns arises with the notion of progress. Keohane mentions, quite dramatically, that

> faced with the governance dilemma, those of us interested in governance on a world scale could retreat to the pure self-interest model. With that set of assumptions, we would probably limit world governance. We would sacrifice gains that could result from better cooperation in order to guard against rule by undemocratic, self-serving institutions responsive, in opaque ways, to powerful elites. If we were successful, the result would be to limit global governance, even at the expense of greater poverty and more violent conflict. We might think ourselves wise, but the results would be sad. Due to excessive fear, we would have sacrificed the liberal vision of progress. (2001: 7)

This implies that political scientists can or will agree on at least four assumptions: (1) that progress is possible, (2) what "progress" means, (3) that we can measure progress, and (4) that we can *trust* whatever metrics we use to assess those forms of progress. Such academic agreement may indeed be not only possible but present. Yet several recent developments and the academic responses (or lack thereof) to these have made me quite skeptical of the notion of "progress" and, in turn, the capabilities of the academy to construct these progressive global institutions along the lines that Keohane suggests.

To wit, we do have several institutions that serve to evaluate the development or stability of countries toward some of the "ideals" that Keohane hints at, such as liberal democracy. Freedom House and Polity are two organizations that routinely evaluate the democratic levels of countries, but what exactly are they measuring? Again I reference the recent U.S. practice of torture. Following the 9/11 attacks, the United States began practicing torture on a limited but notable basis with both foreign detainees and, on an even more limited (but also, I would add, more notable) basis, against at least one U.S. citizen (Jose Padilla) on U.S. soil. Apparently even these developments could not affect the "democratic" score each agency assigned to the United States during this time.[3] So be it, but in my view, any study that uses either of these ranking data sets as a basis on which to study "democratic peace" hereafter (and, judging by its popularity as a pseudoscientific research field, there will be many more of these studies) should be required to note how the rankings elide the institutionalization of coercive interrogations in certain "fully formed" democracies like the United States. Such a disclosure, to me, would constitute "progress" in the field of IR.

---

3. See also Kratochwil's (2006) identification of the value-laden nature of these rankings. Freedom House, to its credit, dealt with this issue by commissioning a panel and then publishing a book based on the panel's findings, titled *Today's American: How Free?* Its conclusions, however, are cause for some concern. While the study expressed dismay over "the restrictions on freedom posed by counterterrorism measures," it concluded, "However, the controversy surrounding the Bush administration's counterterrorism policies must be assessed at least in part within the context of the strains on civil liberties in previous times of war." The study then cited several examples, all of which illustrated much worse transgressions of individual liberty, from these previous periods. The study then stated, "The excesses of these earlier periods of conflict were each addressed and resolved by the normal workings of the American system, though often after some delay. Regarding today's civil liberties controversies, a process of rethinking and adjustment is already under way as this work goes to print." Unfortunately, what this failed to consider was how those previous periods were all cases of *terminable* conflict against a largely identifiable threat, unlike the War on Terror. See "Overview Essay," http://www.freedomhouse.org/template.cfm?page=384 &key=44&parent=5&report=61.

"The Hubris engendered by . . . claims to political wisdom will likely lead to disaster"

Even with the previously noted concerns, we can still find room for an analytics that accepts the variability and spontaneity of life, its ambiguities and complexities, without recourse to either a nihilism or a heroic assembly of global institutions. Vibeke Schou Tjalve titles this "epistemological skepticism" vis-à-vis our "capacity to attain universal rules of coexistence" (2008: 67), or the acknowledgment that in our localized setting, we cannot ever attain the universal. Instead, as she suggests in her reading of Reinhold Niebuhr, we must "learn to live responsibly with finitude, enabling the creative dimensions of human vitality to thrive, while limiting its potential for destruction" (ibid.).

How do we thread such a needle, though? On the one hand, Michael Williams (2004) casts Hobbes's project as an attempt to create an epistemology that can help to reduce the uncertainty of life. This particular treatment of Hobbes provides a more humble proposal than Keohane's regarding the problem of order, and it also advances, at its core, a principle I turn to in the next section of this conclusion—restraint—which can promote a healthy sense of self-enrichment that stems, if not resolves, the dilemma of aesthetic insecurity. Hobbes entered a world full of uncertainty and anxiety and sought to construct not only a political order that could serve as a bulwark against this but also an epistemology that could served as a foundation to "re-construct the practices of knowledge in order to re-establish principles of obligation and authority—and thus sovereignty—in the era of endemic conflict" (Williams 2004: 20). We must know what counts as knowing, however, so through the construction of concepts that could provide some basis for agreement—an episteme—a sound political order would reduce (but never eliminate) such anxiety to a manageable level.

One way to read Williams's treatment of Hobbes is to see it as moving in the opposite direction of the counterpower account I have provided in the previous chapters. Hobbes was concerned about "epistemological indeterminacy," and so, rather than assuming rationality a priori, Hobbes's "Wilful Realism" sought to create the basis within which rational actors could be produced (Williams 2004: 20). Counterpower instead assumes an actor constantly creating its own Self *without* and even against societal codes, because "one cannot specify the nature of [an] aesthetics of existence in advance, any more than actual aesthetic norms and practices are determined by theories of the beautiful or the sublime" (Osborne 1999: 48).

Yet we might also read Williams's treatment of Hobbes in a much more complementary tone, for it intersects with an account of counter-power in two fashions. The first is the need to construct a *pragmatic* ethos, an admission not that "truth" exists but that from the construction of *a* truth, we can establish some basis for order and thus individual creativity. Hobbes creates a sovereign that checks itself, that recognizes its own limitations, a *disciplined* Self that "confronts its own desires and vanity in order to limit them and thereby create a realm of freedom and order in which a maximum pursuit of individuality becomes possible" (Williams 2004: 39). The counterpower practice of parrhesia can help bring this about, as portions of chapter 3 suggested.

Out of the notion of limits comes a second intersection—a corollary commitment to skepticism regarding the very episteme that the sovereign establishes. A healthy amount of such skepticism "leads to a suspicion of, and attack against, dogmatism and (in Hobbes's sense) irrationalism" (Williams 2004: 44). Counterpower, by refracting the Self of power on itself, can help provide the "shock" necessary for such skepticism. It shows the sovereign its own aesthetic limits without seeking to transcend them.

The product of such intersections is quite counterintuitive here, for once we recognize that we cannot do everything or even most anything, we create a healthy basis to realize our own agency. By establishing boundaries of self-operation, we are no longer foiled by the tragedy of hubris, by the prison of "limitless possibilities," but instead are freed, within our own (in part) *chosen* limits, to act as we wish, rather than trying to ascertain a timeless "political wisdom" that constantly eludes us. In order to get there, we need to recognize that we have this capacity to establish limits, to restrain ourselves.[4] Restraint is therefore the primary principle we might turn toward to begin settling with the dilemma of the insecurity that results from the Self's need to create.

"Direct the gaze upon ourselves": Toward the Restraint of the Aesthetic Self

While I think it proves difficult to ever control aesthetic vulnerability, practices that can better restrain the aesthetic Self may serve to diminish the type of ontological "shock" that counterpower engenders. Recall the critique of vitalism in chapter 1 and some of the examples that were re-

---

4. For more on the theme of what I once titled the "recognizing [of] limits and the possibility of freedom," see Steele 2007b: 281–83.

viewed in several of the chapters in this book. Through these examples, we see the Self's need to seek out an action that nevertheless reveals a real Self that starkly and negatively diverges from the idealized one inherent in the narcissism of power. The rhythmic, psychological, and imaginative strata of power's subjectivity are briefly but intensely damaged in this revelation. To reduce such occurrences, the Self needs to be constantly interrogated prior to the revelatory moment, "cut down to size" in its plans and stated purposes. In other words, if the Self becomes damaged in part because of a need to reveal (and revel in) its physique and if the limits that the Self breaks through are in part shaped by its own movements, then securitization of the Self follows from knowing its limits. Security is facilitated through restraint.

There are a surprising variety of theorists we can consult to help shape such an ethos, many of which were discussed in earlier chapters of this book. For Michel Foucault, certain Stoic philosophers (such as Seneca) provided insights, seeing that the organization of the Self begins with organizing knowledge in terms of a *tekhne tou biou,* an "art of living" (2005: 259–60). Working on the Self in a healthy way occurs through four precepts. The first of these is to defeat one's own vices, what Foucault titles *self-control.* The notion that we can control the Self is as liberating as it is depressing—for if we can control our urges, we are therefore responsible for our actions. This becomes especially difficult when we face a crisis, where the Stoics bring us to our second precept: "to be steadfast and serene in the face of adversity and misfortune" (ibid., 265). Adversity is an integral part of the human condition and of our environment (social or natural). Knowing this requires that we calm down and try to maintain as much autonomy in the face of a crisis as possible. A third and related precept is to struggle *against* pleasure—instead of seeking it out, we should resist that which tempts our aesthetic needs. Foucault titles self-control an "internal struggle," whereas steadfastness and struggling against pleasure are "external" struggles. The fourth precept (and final one for our purposes)[5] is to search for happiness within, "in ourselves, in our minds, in the quality of our souls" (ibid.), rather than seeking such life satisfaction through a necessarily societal or communal set of objective criteria (money, fame, fortune, property, materials, etc.).

We might also recognize how much these precepts, in their collective

---

5. There are actually five Stoic precepts Foucault addresses. The fifth Stoic precept is to "be free to depart, to have the soul on our lips." In other words, having resolved a relationship to the Self, we should then be ready for and even rejoice in dying.

promotion of restraint, overlap with the assertions made by other groups of scholars. Classical realists such as Morgenthau, Niebuhr, Wolfers, and Herz, among others, collectively shepherded a normative "prudence" of statecraft and preached what Alberto Coll rightly identifies as a "Burkean catalogue of character traits," which included "the exercise of self-control against the passions and delusions of the mind; profound skepticism toward any attempt to turn theoretical conclusion or 'universals' directly into policy without due regard for the friction of circumstances or 'particulars'" (1996: 92). In a previous work, I have classified the emphasis on self-control in and of itself (internal struggle) and especially in the face of adversity as a principle of "prudence-as-stoicism" (2007b: 279). When faced with a crisis, prudence-as-stoicism resists the urge to "demonstrate" power to an adversary (the supposed source of adversity). Power here exists in self-discipline, in a different form of "demonstration of the will." It is a poker face, an expression of power without expression, one that comes via the (cap)ability to conceal.[6]

However intuitive it may appear, the notion of restraint in the face of adversity is not an easy one to make in the public sphere of U.S. politics in a post-9/11 world. As Niebuhr noted over seventy-five years ago, the public's "rational understanding of political issues remains such a minimum force that national unity of action can be achieved only upon projects initiated by dominant groups . . . or supported by the popular emotions and hysterias which from time to time run through a nation" (1932: 59). It is hardly in a democratic leader's interest to quell such hysteria when it occurs. Yet this is why, I think, scholars studying public opinion have also expressed increasing concern since 9/11 with the way in which U.S. leaders have handled the increase in societal anxiety that followed those attacks. John Mueller's (2005) study, for instance, analyzed the tendency for U.S. policymakers to overemphasize and overreact to a variety of threats over time. Such alarmism is particularly costly against a terrorist threat that thrives on the Self's hysterical work on itself. Thus, Mueller proposes that a series of catastrophes where tens of thousands of Americans died in an attack could be absorbed and that, in words that I

---

6. Jace Weaver notes that it was the *appearance* of the United States in Vietnam that transformed Niebuhr into a full-fledged skeptic of the irony of American culture. In Vietnam, according to Weaver's read of Niebuhr, "the United States looked less like a mature, responsible world power than it resembled a giant, deranged postal employee, who, having shot everyone in sight, turned the gun on itself as well" (1995: 242). In exposing the U.S. Self to an unnecessary excursion into Southeast Asia, the U.S. appeared even more vulnerable than it would have had it not pursued any actions at all.

think go to the heart of my critique of vitalism, "the only way [terrorism] could 'do away with our way of life' would be if *we did that to ourselves in a reaction*" (Mueller 2005: 225, emphasis added).

The problem that Mueller notes here also intersects with the thrust of Ole Wæver's (1995) work on securitization and desecuritization: that securitizing a threat engages a process that, once begun, becomes difficult to reign in. Thus, desecuritization requires even more effort than the initial act of securitization. Yet the same nature of a terrorist threat that makes it so unpredictable also creates one possible mechanism for desecuritization. Because the threat is not seen, attacks are not repelled, but, rather, potential attacks are "foiled," "broken up," before they can materialize. We can only know that progress against threats is being made if we take it *on faith* from those in power. The glitch, of course, is when these authorities make utter fools of themselves—as a relatively new concoction of the U.S. Homeland Security Department did when it suggested in February 2003 that Americans protect themselves from chemical, biological, or radiological attacks by wrapping their homes in plastic and duct tape. This, in addition to the "color-coded" threat system, helped make the department and its warnings an object of ridicule for many Americans and provided much fodder for comedians such as Lewis Black.[7] Thus, it appears that a "boy who cried wolf" syndrome has helped, on its own, to desecuritize the hysteria that was heightened following 9/11.

The situation that we face today is similar (but not identical) to the one classical realists of the immediate postwar era faced: overreaction of violent power to somewhat ambiguous threats. While the nature of the overreaction in that period was much more frightening than it is today (with nuclear weapons still viable instruments of force), contemporary threats ("terror," environmental, health), though identifiable, are nevertheless more ambiguous than they were back then. Restraint, while difficult in the past, was *more* possible then, via the construction of a narrow view of the national interest or through the metaphor of a balance of power (see Little 2007). When we add the aesthetic, however, we are engaging an even deeper stimulus—and a deeper transgression that is much more difficult to restrain when provoked. When the psychological, rhythmic, and imaginative strata of power are all ruptured, it is a her-

---

7. Black's engagement with the issue occurred in his comedy special *Black on Broadway* (2004). The United States does not, of course, hold a monopoly on such public warning systems. Similar systems are utilized in the United Kingdom ("threat level") and France (Vigipirate).

culean task to resist reacting. This is the challenge of an aesthetics of insecurity.

The Emerging "Aversion to Grandiosity"

For those who accept that the challenge of an aesthetics of insecurity can only be met by a healthy ethos of restraint, there is, I think, an emerging generation of scholars ready to promote the notions of limits, pragmatism, and epistemological skepticism. By doing so, they also resist (if not reject) much of the style contained in proposals like Keohane's noted earlier in this conclusion. To avoid charges of narcissism, let me say that I speak not for this generation but as someone within it—shaped, to return to some of the themes of this book, by particular "formative experiences" that we collectively share. Most difficult for this generation, titled "Generation X" by some, "13ers" by others (Strauss and Howe 1991: 317–34), is the notion that comprehensive progress can be achieved by assessing societal development through a universal set of criteria.[8] Generation X has been termed "ambiguous, random and contradictory" (Stephney 2008). It evaded collective evaluation.

It is all well and good to concoct comprehensive standards of assessment, but when the rubber meets the road and we implement those standards as the posts around which the tomato vines are to develop, we are at the mercy (to extend the metaphor some more) of the gardeners, the weather, the soil, and the seeds themselves. When Keohane discusses which values ideally should embody the emerging institutions of global governance, neither he nor the political scientists who buy into this grand project recognize, I fear, just how much human relations elide control. Institutionalization based on even well-constructed universals still, at the end of the day, constrains creativity, which itself is not so much a contending value but a *pattern* of the Self's work on the Self. Our drive for aesthetic creativity cannot be corralled—it can only be conditioned, at best.

Several studies conducted by scholars both inside and outside of the U.S. academy demonstrate this very skeptical stance toward comprehensive solutions or institutions, so my sense is that such a generational *style*,

---

8. I discuss the intersections with classical forms of functionalism later in this conclusion, but David Mitrany's quote regarding "progress" is most instructive here: "In this awkward field we cannot make progress by propounding schemes which have a pleasant symmetry without regard to the rough and shifty terrain on which they have to be grounded" (1977: 70).

far short of a coordinated critique or wave of resistance, nevertheless knows no geographic boundaries.[9] Take, for instance, Cian O'Driscoll's very astute critique of Jean Bethke Elshtain's position that the U.S. War on Terror can be understood using the precepts of Just War doctrine. O'Driscoll proposes that Elshtain lacks a "hermeneutics of suspicion" and, instead, is "too quick to place her faith in American power" (2007: 489).[10] We can also seek counsel from the work of Oded Löwenheim, discussed in chapter 3, on the disciplining nature of "rating and ranking" institutions like Freedom House and Transparency International. Löwenheim configures these not as institutional metrics of progress but, instead, as "examinations" that "legitimize and lead to a more demanding, uncompromising, and intolerant policy or attitude toward the rated/ranked actor, who is now 'known' in comparison to other examinees in an ostensibly exact and unbiased manner" (2007b: 7). In another example, Eric Heinze helpfully terms recent U.S. policies, including the Iraq War, as a "New Utopianism" preoccupied with "what the world *ought* to resemble rather than confronting the realities of the world *as it is*" (2008: 118–19).

None of these works impugn the *motives* of their targets, whether the target is an established academic-intellectual, a philosophy, or a high-level American official. In fact, in all of these cases, we could take each of the targets at their word that they possessed good intentions. It is not intentions this generation is seeking to reveal. To follow Patrick Jackson, "sincerity is difficult enough to evaluate in private life," and "trying to determine the sincerity" of these individuals seems "nigh upon impossible" (2006: 22). Rather, the emerging generation seeks to expose the previous generation's propensity for the *grandiose*. These skeptical Gen X scholars critique the notion that comprehensive solutions can ever be discovered, whether in the form of the Just War doctrine or a democratic ideal of political development. Their works contain stylistic abilities to "pour ice-water realism" on another generation's audacious plans (Strauss and Howe 1991: 251), and they all do so by returning to what Tjalve characterizes as a more "genuine realism" that "holds power to be a permanent source of temptation. [For] today, our 'realist' strategies are concerned with designing the future of the globe. If that is not hybris, then what is?" (2008: 143). By contrast, the style of this emerging

9. This follows from the form (if not the policy-advising function) of epistemic communities, first detailed in Haas (1989) as transnational collaborative networks of scholars.
10. Vibeke Shou Tjalve's previously mentioned study concludes (2008: 140–42) with a similar critique of Elshtain's work.

generation of scholars speaks of an "incorrigible aversion to grandiosity" that was at the center of another "reactive" generation—the so-called Lost Generation of Niebhur, Hemingway, and Bourne (Strauss and Howe 1991: 251).[11]

Acupunctural Formations

There is little good news to report to those scholars who share a concerned sense that the field of IR is careening "toward what seems to be an increasing tribalism" (Vertzberger 2005: 120). That said, the prudential acceptance that a field of power relations will never be transcended and that an emerging generation of scholars may be content with this nevertheless also allows us—*frees us*—to move from seeking out transcendent panaceas toward the temporary, localized, but nevertheless *attainable* possibilities that contain opportunities for productive human relations.[12]

Combined with an ethos of restraint, which sketches the contours of a limited field within which the Self operates, acupunctural formations are those practices that occur in very localized settings, with the possibility (but not inevitability) that if this localized area is "pierced," it can "engender a ripple effect" (Gordinier 2008: 150). Like counterpower, acupunctural formations are spontaneous, but unlike counterpower, which disturbs, acupunctural formations, like the medical treatment used as its metaphor, heal (but never "cure"). Such formations do not ignore the macroforces that produce common problems. They instead slice into such problems at a microlevel.

For an example, we can turn to the approach to housing blight that has taken hold in the form of "urban acupuncture," a practice defined in one recent study as "the selective redevelopment of appropriate sites within the historic fabric" (Eames et al. 2007: 2). The key word here is *appropriate,* as acupunctural formations, because they exist at the local and immediate level, serve particular, pragmatic purposes. An approach to acupunctural formations forever surrenders the idea that such mi-

---

11. Although not born in the United States (Romanian born and then a British citizen), the aforementioned father of classical functionalism, David Mitrany, was also born during the same epoch as the Lost Generation cohort (two years after Bourne and four years before Niebuhr).

12. As David Campbell notes, ethical relations may only be possible when we resist a "prior normative framework," or "being 'against' theory, ethics and justice [is] an affirmative position designed to foster the ethical relation" (2000: 109).

crostructures can be extrapolated into comprehensive solutions. Like the process of restraint, acupunctural formations are celebrated through an ethos of limitation, one that embraces the eclectic, the individualized, and the unique. Acupunctural formations elide generalization in part because they cannot work when they are mass-produced or genericized. The incentive that brings them about as spontaneous constructions is ruined when they are normalized into a cliché.[13] In other words, that which comes about organically loses its effectiveness when it no longer exudes innovation.

In this vein, those who study acupunctural formations may find the early classical functionalist work, especially that of David Mitrany (mentioned briefly in notes earlier in this conclusion), to be quite useful. Before it was systematized into its "neo"-scientific form during the behavioral revolution, this functionalism embodied an acupunctural ethos. Localized and pragmatic, it was captured quite well, in my view, by Mitrany, in a reflective autobiographical sketch, as getting up "each day [looking] for fair answers to questions[,] almost every one of which was too tangled to allow a straight answer" ([1969] 1975: 10). This functionalism did not attempt to solve every problem out there but took the complexity of global politics as a fact. As opposed to solutions "from above," such "functional arrangements have the virtue of technical self-determination . . . The nature of each function tells of itself the scope and powers needed for its effective performance" (Mitrany 1977: 73). Notice the inherent aesthetic activity in such functionalist arrangements—we wake up each day, free of systematic plans, and are free, even forced, to create in the here and now. An acupunctural attitude results from the need "to do something that matters, if only on a small scale" (Strauss and Howe 1991: 333). Because we have no universal model to "evaluate" our work, we can create it free of expectations and constrictive standards. Such arrangements are a "ground zero" where, to paraphrase Mitrany, the "realities of everyday life pulsate" (1977: 74).

---

13. In a note in the introduction to this book, I briefly mentioned the novel *The Unbearable Lightness of Being* by Milan Kundera in the context of discussing counterpower. Kundera's development of the concept of totalitarian kitsch is similar to the notion of cliché. Kundera likens kitsch to an inquisition; others view kitsch as the opposite of authentic, as "fake" (Shiner 2001: 271) or as "bad taste" (Ward 1994). There is a generational element here, too, according to Jeff Gordinier, who writes that if there is one "thing that unites [Generation] Xers in scorn," it is kitsch: "Xers bristle when they're told how to vote, how to behave, what to listen to, how to squander their time. They recoil at any hint of a presumption that *this is how things are done* or, even worse, *this is what you're supposed to think*" (2008: 47).

Additionally, modern technology has provided, as I mentioned at the end of chapter 4, the means through which aesthetic power will be perpetually insecure through counterpower images. Yet we need not resist aesthetic power to recognize how it can work back in a productive manner on we who stand within it—because the folly of aesthetic power can be very entertaining. Orwell's response to why "the goose-step is not used in England" is instructive here: "It is not used because the people in the street would laugh. Beyond a certain point, military display is only possible in countries where the common people dare not laugh at the army" (1941). We are never submitting to power if we still contain the capacity to look upon its comedic elements—if we can, moreover, ridicule this comedy and revel in its ugliness. Against this ugliness, we all contain a beauty that can never be extinguished. Power will always be vulnerable as long as we have this capability to create, to not evaluate it based on our fear (or awe) of when it is practiced but judge it aesthetically.

The ability to laugh is another acupunctural formation because it contains in it the primary benefit of acupuncture—the capacity to heal. Yet just like the beauty of urban acupuncture or the flexible benefits of functional arrangements, this is a spontaneous and unplanned event that will not cure us or emancipate us from a field of operations. Acupuncture is used to *treat,* rather than cure, permanent conditions that humans face. There is nothing heroic per se about acupunctural formations. But at the end of the day, is this not how we should face power—recognizing that it is always with us but that it never need direct us in our every action? By acupuncturing a formation within power, we reveal our human capacities. Such capacities give us the possibility to do the unpredictable.

Thus, we can construct an analytical attitude to accompany the ethos of acupuncture that *embraces* this unpredictability and uncertainty of human relations. Such an analytical attitude, if it can be refined, may continue to investigate how the seemingly confident, orderly, and secure in global politics are anything but. Indeed, such security may be, like the goose step, a synchronized facade. This analytic attitude might move toward recognizing the aesthetics of such insecurity, the beauty of power's vulnerability, and might even become cautiously comfortable with embracing the moment when such power is engaged, to recognize the calm before the inevitable reactive storm that will follow, the knowledge that *something* "good" or "bad" will ensue because of power's tendencies and proclivities. Francois Debrix calls such occasions "events-as-surprise," which admittedly "have no guarantee of political or democratic success," yet "still, their happening, their presence, and their surprise may nonetheless actualize a critique of ideology and politics and, as such, *force an opening for a freedom or democracy to come*" (2006: 787–88, emphasis added). Such an attitude that "seeks to restore the quality of surprise to the event . . . may also be open to differential possibilities for democracy in contemporary geopolitics and international relations" (Debrix 2007: 127). We do not know from when or where these events may originate, but neither does an organized body of power, and it is there where the beauty of uncertainty lies. That this uncertainty can never be once and for all resolved should not lead us to despair, however, as long as we can turn such uncertainty into a healthy form of epistemological skepticism, as evidenced by the Hobbesian project that Williams proposed.

Foucault's aesthetics of existence, of course, provides a further logical basis for such an attitude, since the notion of innovation presupposes uncertainty, unpredictability, or, as Osborne avers, "the product of an on-going experiment or process of trial and error . . . the model of the artist, who, on the basis of a hard ascetic labor, enters into the *unknown*" (1999: 48, emphasis added). But we can also consult, for a final occasion (in this book, anyway), the words of Hans Morgenthau, who wrote during a 1954 conference discussing the merits of a "theory of international relations" that such a theory "must guard, then, against the temptation to take itself too seriously and to neglect the ambiguities which call it into question at every turn" (quoted in Guilhot 2008: 297). Social life is so complex that we who study it are bound to mistake its operation, so we should be humble in its presence and never believe we can contain it because we have stumbled on some eternal truths. There will be, then, to paraphrase Richard Bernstein's (1983: 18) oft-quoted phrase (cf. Campbell 1998: 193), no "stable rock upon which we can secure our lives against the vicissitudes that constantly threaten us," or, in the case of studying the art of counterpower, the vicissitudes that insolventize power's ontology. An analytical attitude that can embrace the human capacity for innovation, creativity, and style may do so with the cautious expectation that the aesthetic need for power to become "beautiful" once again also serves as the basis to, even slightly, make it recognize its *own* capacity to transform.

References

Abrahamsen, Rita, and Michael C. Williams. 2007. Securing the City: Private Security Companies and Non-state Authority in Global Governance. *International Relations* 21:237–53.

Ackerly, Brooke, and Jacqui True. 2008. Reflexivity in Practice: Power and Ethics in Feminist Research. *International Studies Review* 10, no. 4: 693–707.

Acuff, Jonathan. 2008. Identity and Legitimacy. PhD diss., University of Washington.

Adorno, Theodore. 1973. *Negative Dialectics*. New York: Seabury Press.

Adorno, Theodore. 1997. *Aesthetic Theory*. Minneapolis: University of Minnesota Press.

Agency France Presse. 2005. U.N. Aid Official Smooths Feathers Ruffled in Washington over "Stingy" Remark. 2 January.

Allen, A. 2002. Power, Subjectivity, and Agency: Between Arendt and Foucault. *International Journal of Philosophical Studies* 10, no. 2: 131–49.

Allison, Graham. 1969. Conceptual Models and the Cuban Missile Crisis. *American Political Science Review* 63, no. 3: 689–718.

Anderson, David L., ed. 1998. *Facing My Lai: Moving beyond the Massacre*. Lawrence: University Press of Kansas.

Arendt, Hannah. [1964] 2006. *Eichmann in Jerusalem*. Repr. New York: Penguin.

Arendt, Hannah. 1969. *On Violence*. New York: Harcourt, Brace, and World.

Aron, Raymond. 1966. *Peace and War: A Theory of International Relations*. Garden City, NY: Doubleday.

Arrigo, Jean Maria. 2004. A Utilitarian Argument against Torture. *Science and Engineering Ethics* 10, no. 3: 543–72.

Avant, Deborah. 2005. *The Market for Force: The Consequences of Privatizing Security*. Cambridge: Cambridge University Press.

Bacevich, Andrew. 2005. *The New American Militarism: How Americans Are Seduced by War*. New York: Oxford University Press.

Bacevich, Andrew. 2007. *The Limits of Power: The End of American Exceptionalism*. New York: Basic Books.

Bach, Jonathan. 1999. *Between Sovereignty and Integration: German Foreign Policy and National Identity after 1989*. New York: St. Martin's Press.

Bai, Matt. 2008. The McCain Doctrines. *New York Times Magazine*, 18 May.

Ballard, John R. 2006. *Fighting for Fallujah*. Westport, CT: Greenwood.

Barnett, Michael N., and Emanuel Adler, eds. 1998. *Security Communities.* Cambridge: Cambridge University Press.

Barnett, Michael N., and Raymond Duvall. 2005. Power in International Politics. *International Organization* 59 (Winter): 39–75.

Barnett, Michael N., and Martha Finnemore. 2004. *Rules for the World.* Ithaca: Cornell University Press.

Baudrillard, Jean. [1981] 1995. *Simulacra and Simulation.* Trans. Sheila Glaser. Ann Arbor: University of Michigan Press.

Beck, Ulrich. 2006. *Power in the Global Age.* Cambridge: Polity.

Beck, Ulrich, and Elisabeth Beck-Gernsheim. 2008. Global Generations and the Trap of Methodological Nationalism for a Cosmopolitan Turn in the Sociology of Youth and Generation. *European Sociological Review* 25, no. 1: 25–36.

Behnke, Andreas, and Linda S. Bishai. 2007. War, Violence, and the Displacement of the Political. In *The International Political Thought of Carl Schmitt. Terror, Liberal War and the Crisis of Global Order,* ed. Louiza Odysseos and Fabio Petito. London: Routledge.

Bennett, Jane. 1996. "How Is It, Then, That We Still Remain Barbarians?" Foucault, Shiller, and the Aestheticization of Ethics. *Political Theory* 24, no. 4: 653–72.

Bennett, Scott, and Allan C. Stam III. 1998. The Declining Advantages of Democracy. *Journal of Conflict Resolution* 42, no. 3: 344–66.

Bennett, Scott D., and Allan C. Stam III. 2000. Research Design and Estimator Choices in the Analysis of Interstate Dyads: When Decisions Matter. *Journal of Conflict Resolution* 44, no. 5: 653–85.

de Benoist, Antoine. 2007. Global Terrorism and the State of Permanent Exception: The Significance of Carl Schmitt's Thought Today. In *The International Political Thought of Carl Schmitt: Terror, Liberal War and the Crisis of Global Order,* ed. Louiza Odysseos and Fabio Petito. London: Routledge.

Bernstein, Richard. 1983. *Beyond Objectivism and Relativism: Science, Hermeneutics, and Praxis.* Philadelphia: University of Pennsylvania Press.

Bially Mattern, Janice. 2005. *Ordering International Politics.* New York: Routledge.

Bially Mattern, Janice. 2007. Why "Soft Power" Isn't So Soft: Representational Force and Attraction in World Politics. In *Power in World Politics,* ed. Felix Berenskoetter and M. J. Williams, 1–22. New York: Routledge.

Bleiker, Roland. 2000. *Popular Dissent, Human Agency and Global Politics.* Cambridge: Cambridge University Press.

Bleiker, Roland. 2001. The Aesthetic Turn in International Political Theory. *Millennium: Journal of International Studies* 30, no. 3: 509–33.

Bleiker, Roland, and Martin Leet. 2006. "From the Sublime to the Subliminal": Fear, Awe, and Wonder in International Politics. *Millennium: Journal of International Studies* 34, no. 3: 713–37.

Bolton, John. 2008. *Surrender Is Not an Option.* New York: Threshold.

Booth, Ken. 1999. Three Tyrannies. In *Human Rights in Global Politics,* ed. Tim Dunne and Nicholas Wheeler, 31–70. Cambridge: Cambridge University Press.

Booth, Ken, and Nicholas Wheeler. 2008. *Security Dilemma: Fear, Cooperation, and Trust in World Politics.* New York: Palgrave.

Boswell, T., and M. Sweat. 1991. Hegemony, Long Waves, and Major Wars: A Time Series Analysis of Systemic Dynamics, 1496–1967. *International Studies Quarterly* 35, no. 2: 123–50.

Bousquet, Antoine. 2006. Time Zero: Hiroshima, September 11 and Apocalyptic Revelations in Historical Consciousness. *Millennium: Journal of International Studies* 34, no. 3: 739–66.

Bradbury, Stephen 2005. Memorandum for John R. Rizzo, Senior Deputy Counsel, Central Intelligence Agency. United States Department of Justice, Office of Legal Counsel, 10 May.

Brooks, Stephen G., and William Wohlforth. 2005. Hard Times for Soft Balancing. *International Security* 30, no. 1: 72–108.

Brown, Chris. 1999. Universal Human Rights: A Critique. In *Human Rights in Global Politics,* ed. Tim Dunne and Nicholas Wheeler, 103–27. Cambridge: Cambridge University Press.

Browne, Malcolm. 1975. *New York Times,* 22 April, 1.

Bruggemeir, Franz-Josef. 2005. *How Green Were the Nazis?* Athens: Ohio University Press.

Brunsson, Nils. 2002. *The Organization of Hypocrisy: Talk, Decisions and Action in Organizations.* Oslo: Copenhagen Business School Press.

Buger, Christian, and Frank Gadinger. 2007. Reassembling and Dissecting: International Relations Practice from a Science Studies Perspective. *International Studies Perspectives* 8:90–110.

Bukovansky, Mlada. 2005. Hypocrisy and Legitimacy: Agricultural Trade in the World Trade Organization. Paper presented at the International Studies Association Annual Meeting, Honolulu, Hawaii.

Bull, Hedley. 1972. The Theory of International Politics, 1919–1969. In *The Aberystwyth Papers: International Politics 1919–1969,* ed. B. Porter. London: Oxford University Press.

Bull, Hedley. 1977. *The Anarchical Society.* New York: Columbia University Press.

Burnham, Walter Dean. 1970. *Critical Elections and the Mainsprings of American Politics.* New York: W. W. Norton.

Buzan, Barry. 1993. From International System to International Society: Structural Realism and Regime Theory Meet the English School. *International Organization* 47, no. 3: 327–52.

Buzan, Barry. 2004. *From International to World Society? English School Theory and the Social Structure of Globalization.* Cambridge: Cambridge University Press.

Buzan, Barry, and Ole Wæver. 2003. *Regions and Powers.* Cambridge: Cambridge University Press.

Bybee, Jay S., and John Yoo. [2002] 2005. Potential Legal Constraints Applicable to Interrogations of Persons Captured by U.S. Armed Forces in Afghanistan. In *The Torture Papers,* ed. Karen Greenberg and Joshua Dratel, 144–71. New York: Cambridge University Press.

Campbell, David. [1992] 1998. *Writing Security.* Repr., Minneapolis: University of Minnesota Press.

Campbell, David. 2000. Justice and International Order: The Case of Bosnia and Kosovo. In *Ethics and International Affairs: Extent and Limits,* ed. Jean-Marc Coicaud and Daniel Warner, 103–27. Tokyo: United Nations University Press.

Castells, Manuel. 2007. Communication, Power and Counterpower in the Network Society. *International Journal of Communication* 1:238–66.

Cederman, Lars-Erik. 2001. Back to Kant: Reinterpreting the Democratic Peace as a Macrohistorical Process. *American Political Science Review* 95, no. 1: 15–31.

Clark, Ian. 2007. *International Legitimacy and World Society.* New York: Oxford University Press.

Coll, Alberto. 1996. Normative Prudence as a Tradition of Statecraft. In *Ethics and In-*

ternational Affairs, ed. Joel H. Rosenthal, 75–100. Washington, DC: Georgetown University Press.

Cooley, Alexander. 2005. *Logics of Hierarchy.* Ithaca: Cornell University Press.

Cornwell, Robert. 2004. Atrocity in Fallujah May Be Turning Point for US. *Independent,* 2 April.

Cox, Robert. 1983. Gramsci, Hegemony, and International Relations: An Essay in Method. *Millennium: Journal of International Studies* 12, no. 2: 162–75.

Cox, Robert. 1986. Social Forces, States and World Orders; Beyond International Relations Theory. In *Neorealism and Its Critics,* ed. Robert Keohane, 204–54. New York: Columbia University Press.

Cox, Robert. 1992. Multilateralism and World Order. *Review of International Studies* 18:161–80.

Cox, Robert. 1999. Civil Society at the Turn of the Millennium: Prospects for an Alternative World Order. *Review of International Studies* 25, no. 1: 3–28.

Cozette, Murielle. 2008. Reclaiming the Critical Dimension of Realism: Hans J. Morgenthau on the Ethics of Scholarship. *Review of International Studies* 34:5–27.

Crawford, Neta. 2000. The Passion of World Politics. *International Security* 24, no. 4 (Spring): 116–56.

Crawford, Neta. 2002. *Argument and Change in World Politics: Ethics, Decolonization, and Humanitarian Intervention.* Cambridge: Cambridge University Press.

Cruz, Consuelo. 2000. Identity and Persuasion: How Nations Remember Their Pasts and Make Their Futures. *World Politics* 52, no. 3: 275–312.

Dahl, Robert. 1957. The Concept of Power. *Behavioral Science* 2, no. 3: 201–15.

Dauber, Cori. 2001. Image as Argument: The Impact of Mogadishu on U.S. Military Intervention. *Armed Forces and Society* (Winter): 205–29.

Davidson, Jenny. 2004. *Hypocrisy and the Politics of Politeness: Manners and Morals from Locke to Austen.* Cambridge: Cambridge University Press.

Debrix, Francois. 1999. *Re-envisioning Peacekeeping: The United Nations and the Mobilization of Ideology.* Minneapolis: University of Minnesota Press.

Debrix, Francois. 2006. The Sublime Spectatorship of War: The Erasure of the Event in America's Politics of Terror and Aesthetics of Violence. *Millennium: Journal of International Studies* 34, no. 3: 767–92.

Debrix, Francois. 2007. *Tabloid Terror: War, Culture and Geopolitics.* London: Routledge.

DeLaet, Debra. 2006. Gender Justice: A Gendered Assessment of Truth-Telling Mechanisms. In *Telling the Truths,* ed. Tristan Anne Borer, 151–80. Notre Dame: University of Notre Dame Press.

Delehanty, Will K., and Brent J. Steele. 2009. Engaging the Narrative in Ontological (In)Security Theory: Insights from Feminist IR. *Cambridge Review of International Affairs* 22, no. 3: 523–40.

der Derian, James. 1990. The (S)pace of International Relations: Simulation, Surveillance, and Speed. *International Studies Quarterly* 34:295–310.

der Derian, James. 2001. *Virtuous War: Mapping the Military-Industrial-Media-Entertainment Network.* Boulder: Westview Press.

der Derian, James. 2002. Cyberspace as Battlespace: The New Virtual Alliance of the Military, the Media and the Entertainment Industry. In *Living with Cyberspace: Technology and Society in the 21st Century,* ed. John Armitage and Joanne Roberts. New York: Continuum.

Deudney, Daniel. 2007. *Bounding Power: Republican Security Theory from the Polis to the Global Village.* Princeton: Princeton University Press.

Dewey, John. 1917. In a time of national hesitation. *Seven Arts Magazine* 2:3–7.

Dewey, John. 1934. *Art as Experience.* New York: Putnam.

DiMaggio, Paul J., and Walter W. Powell. 1983. The Iron Cage Revisited: Institutional Isomorphism and Collective Rationality in Organizational Fields. *American Sociological Review* 48, no. 2: 147–60.

Doty, Roxanne Lynn. 1996. *Imperial Encounters: The Politics of Representation in North-South Relations.* Minneapolis: University of Minnesota Press.

Drulak, Petr. 2006. Reflexivity and Structural Change. In *Constructivism and International Relations,* ed. Stefano Guzzini and Anna Leander. New York: Routledge.

Duffy, Trent. 2004. Press gaggle. http://www.whitehouse.gov/news/releases/2004/12/20041228-2.html.

Dunne, Timothy. 1997. Colonial Encounters in International Relations: Reading Wight, Writing Australia. *Australian Journal of International Affairs* 51, no. 3: 309–23.

Dunne, Timothy, and Nicholas Wheeler. 1999. *Human Rights in Global Politics.* Cambridge: Cambridge University Press.

Eames, Brian, et al. 2007. Urban Acupuncture—A Methodology for the Sustainable Rehabilitation of "Society Buildings" in Vancouver's Chinatown into Contemporary Housing. http://dsp-psd.pwgsc.gc.ca/collection_2008/cmhc-schl/nh18-23/ NH18-23-107-008E.pdf.

Eckhardt, William. 2000. My Lai: An American Tragedy. http://www.law.umkc.edu/faculty/projects/ftrials/mylai/ecktragedy.htm.

Edkins, Jenny. 1999. *Postructuralism and International Relations: Bringing the Political Back In.* Boulder: Lynne Rienner.

Edkins, Jenny. 2002. Forget Trauma? Responses to September 11. *International Relations* 16, no. 2: 243–56.

Edkins, Jenny. 2003. *Trauma and the Memory of Politics.* Cambridge: Cambridge University Press.

Elshtain, Jean Bethke. 2004. *Just War against Terror.* New York: Basic Books.

Elster, Jon. 1996. Strategic Uses of Argument. In *Barriers to Conflict Resolution,* ed. K. J. Arrow et al. New York: W. W. Norton.

Enloe, Cynthia. 2007. Feminism. In *International Relations Theory for the Twenty-First Century: An Introduction,* ed. Martin Griffiths. London: Routledge.

Feith, Douglas. 2008. *War and Decision.* New York: HarperCollins.

Fierke, Karin M. 2007. *Critical Approaches to International Security.* Cambridge: Polity.

Filson, Darren, and Suzanne Werner. 2004. Bargaining and Fighting: The Impact of Regime Type on War Onset, Duration, and Outcomes. *American Journal of Political Science* 48, no. 2: 296–313.

Finnemore, Martha. 2003. *The Purpose of Intervention.* Ithaca: Cornell University Press.

Finney, John. 1975. Minh Offers Unconditional Surrender. *New York Times,* 30 April.

Foucault, Michel. 1970. *The Order of Things.* London: Tavistock.

Foucault, Michel. 1977a. *Language, Counter-Memory, and Practice.* Ed. Donald Bouchard. Trans. Donald Bouchard and Sherry Simon. Ithaca: Cornell University Press.

Foucault, Michel. 1977b. *Discipline and Punish.* Vintage.

Foucault, Michel. 1980. *Power/Knowledge: Selected Interviews and Other Writings: 1972–1977.* Ed. Colin Gordon. Trans. Colin Gordon et al. Brighton: Harvester.

Foucault, Michel. 1982. *The History of Sexuality.* Vol. 2, *The Uses of Pleasure.* New York: Vintage Books.

Foucault, Michel. 1984. *The History of Sexuality*. Vol. 3, *The Care of the Self*. New York: Vintage Books.

Foucault, Michel. 1988. *Politics, Philosophy, Culture*. Ed. Lawrence Kritzman. Trans. Alan Sheridan et al. New York: Routledge.

Foucault, Michel. 1989. *Foucault Live*. Ed. Sylvère Lotringer. Trans. John Johnston. Los Angeles: Semiotext(e).

Foucault, Michel. 1997. Of Other Spaces: Utopias and Heterotopias. In *Rethinking Architecture: A Reader in Cultural Theory*, ed. Neil Leach. London: Routledge.

Foucault, Michel. 2001. *Fearless Speech*. Ed. Joseph Pearson. Los Angeles: Semiotext(e).

Foucault, Michel. 2005. *The Hermeneutics of the Subject*. Ed. Frédéric Gros. Trans. Graham Burchell. New York: Palgrave-MacMillan.

Foucault, Michel. 2007. *The Politics of Truth*. Ed. Sylvère Lotringer. Trans. Lysa Hochroth and Catherine Porter. Los Angeles: Semiotext(e).

Fredrickson, David E.. 1996. Parrhesia in Pauline Epistles. In *Friendship, Flattery and Frankness of Speech*, ed. John Fitzgerald, 163–84. Leiden: Brill.

Fromm, Erich. [1941] 1964. *Escape from Freedom*. Repr. Holt, Rinehart and Winston.

Frost, Mervyn. 1996. *Ethics in International Relations: A Constitutive Theory*. Cambridge: Cambridge University Press.

Garrison, Jim. 1998. Foucault, Dewey, and Self-Creation. *Educational Philosophy and Theory* 30, no. 2: 111–34.

Gartner, Scott Sigmund. 1997. *Strategic Assessment in War*. New Haven: Yale University Press.

Giddens, Anthony. 1984. *The Constitution of Society*. Berkeley: University of California Press.

Giddens, Anthony. 1991. *Modernity and Self-Identity*. Stanford: Stanford University Press.

Gill, Stephen. 1990. *Gramsci and Global Politics: Towards a Post-Hegemonic Research Agenda*. New York: Columbia University Press.

Gill, Stephen, and David Law. 1988. *The Global Political Economy: Perspectives, Problems, and Policies*. Baltimore: Johns Hopkins University Press.

Gilpin, Robert. 1981. *War and Change in International Politics*. Cambridge: Cambridge University Press.

Glad, Clarence E. 1996. Frank Speech, Flattery, and Friendship in Philodemus. In *Friendship, Flattery and Frankness of Speech*, ed. John Fitzgerald, 21–60. Leiden: Brill.

Gleditsch, Nils Peter. 2008. The Liberal Moment Fifteen Years On: Presidential Address, 49th Convention of the International Studies Association. *International Studies Quarterly* 52, no. 4: 691–712.

Goldstein, Joshua. 1988. *Long Cycles*. New Haven: Yale University Press.

Gordinier, Jeff. 2008. *X Saves the World*. New York: Viking Press.

Gould, Harry. 2003. Constructivist International Relations Theory and the Semantics of Performative Language. In *Language, Agency and Politics in a Constructed World*, ed. Francois Debrix, 50–65. New York: M. E. Sharpe.

Gourevitch, Philip, and Errol Morris. 2008. *Standard Operating Procedure*. New York: Penguin.

Greenberg, Karen, and Joshua Dratel. 2005. *The Torture Papers*. New York: Cambridge University Press.

Greenhill, Brian. 2008. Recognition and Collective Identity Formation in International Politics. *European Journal of International Relations* 14, no. 2: 343–68.

Guilhot, Nicolas. 2008. The Realist Gambit: Postwar American Political Science and the Birth of IR Theory. *International Political Sociology* 2:281–304.

Gutmann, Amy, and Dennis Thompson. 2000. The Moral Foundations of Truth Commissions. In *Truth v. Justice: The Morality of Truth Commissions,* ed. Robert I. Rotberg and Dennis Thompson, 22–44. Princeton: Princeton University Press.

Guzzini, Stefano. 2000. A Reconstruction of Constructivism in International Relations. *European Journal of International Relations* 6, no. 2: 147–82.

Haas, Peter M. 1989. Do Regimes Matter? Epistemic Communities and Mediterranean Pollution Control. *International Organization* 43:377–403.

Habermas, Jürgen. 1984. *The Theory of Communicative Action.* Trans. T. McCarthy. Vol. 1, *Reason and the Rationalization of Society.* Boston: Beacon Press.

Habermas, Jürgen. 1990. *Moral Consciousness and Communicative Action.* Cambridge, MA: MIT Press.

Hall, Rodney Bruce. 1997. Moral Authority as a Power Resource. *International Organization* 51:591–622.

Hall, Rodney Bruce. 2008. Deontic Power in International Politics. Manuscript, University of Oxford.

Hampson, F. 2002. *Madness in the Multitude: Human Security and World Disorder.* London: Oxford University Press.

Hanson, Stephen. 1997. *Time and Revolution.* Chapel Hill: University of North Carolina Press.

Harre, Rom. 1998. *The Singular Self.* London: Sage.

Hearn, Alison. 2008. Meat, Mask, Burden: Probing the Contours of the Branded "Self." *Journal of Consumer Culture* 8, no. 2: 197–217.

Heinze, Eric A. 2008. The New Utopianism: Liberalism, American Foreign Policy, and the War in Iraq. *Journal of International Political Theory* 4, no. 1: 105–25.

Hemingway, Ernest. 1929. *Farewell to Arms.* New York: Scribner.

Herborth, Benjamin. 2008. The Idea of Semantic Colonization. Paper presented at the International Studies Association Annual Meeting, San Francisco, CA.

Heritage Foundation. 2004. American Generosity Is Underappreciated. http://www.heritage.org/research/tradeandforeignaid/wm630.cfm.

Hersh, Seymour. 1970. *My Lai 4: A Report on the Massacre and Its Aftermath.* New York: Random House.

Hersh, Seymour. 1972. *Cover-up: The Army's Secret Investigation of the Massacre at My Lai 4.* New York: Random House.

Hersh, Seymour. 2004. Torture at Abu Ghraib. *New Yorker,* 10 May.

Hersh, Seymour. 2007. The Gray Zone. *New Yorker,* 24 May. Available at http://www.newyorker.com/archive/2004/05/24/040524fa_fact.

Herz, John. 1950. Idealist Internationalism and the Security Dilemma. *World Politics* 2, no. 2: 157–80.

Herz, John. 1959. *International Politics in the Atomic Age.* New York: Columbia University Press.

Hoffmann, Stanley. 1996. In Defense of Mother Teresa: Morality in Foreign Policy. *Foreign Affairs* 75, no. 2: 172–75.

Holden, Gerald. 2006. Cinematic IR, the Sublime, and the Indistinctiveness of Art. *Millennium: Journal of International Studies* 34, no. 3: 793–818.

Hom, Andrew. Forthcoming. Hegemonic Metronome: The Ascendancy of Western Standard Time. *Review of International Studies.*

Horner, Charles. 1980. America Five Years after Defeat. *Commentary* (April).

Howell, Alison. 2007. Victims or Madmen? The Diagnostic Competition over "Terrorist" Detainees at Guantanamo Bay. *International Political Sociology* 1, no. 1: 29–48.

Huntington, Samuel. 1991. How Countries Democratize. *Political Science Quarterly* 106, no. 4: 579–616.

Huntley, Wade. 1996. Kant's Third Image: Systemic Sources of the Liberal Peace. *International Studies Quarterly* 40, no. 1: 45–76.

Hutchings, Kimberly. 2007. Time and Critique in IR Theory. *Review of International Relations* 33, no. 1: 71–89.

Hutchings, Kimberly. 2008. 1988 and 1998: Contrast and Continuity in Feminist International Relations. *Millennium: Journal of International Studies* 37, no. 1: 97–105.

Huysmans, Jeffrey. 1998a. Security! What Do You Mean? From Concept to Thick Signifier. *European Journal of International Relations* 4, no. 2: 226–55.

Huysmans, Jeffrey. 1998b. The Question of the Limit: Desecuritization and the Aesthetics of Horror in Political Realism. *Millennium* 27, no. 2: 569–89.

Hymans, Jacques E. C. 2006. *The Psychology of Nuclear Proliferation: Identity, Emotions, and Foreign Policy.* Cambridge: Cambridge University Press.

Ikenberry, G. John. 1999. Institutions, Strategic Restraint, and the Persistence of American Postwar Order. *International Security* 23, no. 3: 43–78.

Ikenberry, G. John. 2002. *America Unrivaled: The Future of the Balance of Power.* Ithaca: Cornell University Press.

Ish-Shalom, Piki. 2006. Theory as a Hermeneutical Mechanism: The Democratic-Peace Thesis and the Politics of Democratization. *European Journal of International Relations* 12, no. 4: 565–98.

Jackson, Patrick. 2003. Defending the West: Occidentalism and the Formation of NATO. *Journal of Political Philosophy* 11, no. 3: 223–52.

Jackson, Patrick T. 2004. Hegel's House or "People Are States Too." *Review of International Studies* 30:281–87.

Jackson, Patrick. 2006. *Civilizing the Enemy: German Reconstruction and the Invention of the West.* Ann Arbor: University of Michigan Press.

Jackson, Patrick. 2008. Foregrounding Ontology. *Review of International Studies* 34, no. 1: 129–54.

Janis, Irving. 1982. *Groupthink.* 2nd ed. Boston: Houghton Mifflin.

Jervis, Robert. 1968. Hypotheses on Misperception. *World Politics* 20, no. 3: 454–79.

Johnson, James. 1993. Is Talk Really Cheap? Prompting Conversation between Critical Theory and Rational Choice. *American Political Science Review* 87, no. 1: 74–86.

Kaplan, Robert. 2002. *Warrior Politics.* New York: Random House.

Kaplan, Sidney. 1956. Social Engineers as Saviors: Effects of World War I on Some American Liberals. *Journal of the History of Ideas* 17:247–69.

Kemal, Salim. 1999. Aesthetic Licence: Foucault's Modernism and Kant's Post-modernism. *International Journal of Philosophical Studies* 7, no. 3: 281–303.

Kennan, George. 1985/86. Morality and Foreign Policy. *Foreign Affairs* (Winter): 205–18.

Kennedy, Paul. 1987. *The Rise and Fall of the Great Powers.* New York: Random House.

Keohane, Robert. 1989. *International Institutions and State Power: Essays in International Relations Theory.* Boulder: Westview Press.

Keohane, Robert. 2000. Ideas Part-way Down. *Review of International Studies* 26, no. 1: 125–30.

Keohane, Robert. 2001. Governance in a Partially Globalized World: Presidential Address, American Political Science Association. *American Political Science Review* 95, no. 1: 1–12.

Kerouac, Jack. [1947] 2007. *On the Road.* 50th anniversary ed. New York: Penguin.

Kessler, Oliver. 2008. The Discourse on Discourses. Paper presented at the International Studies Association Annual Meeting, San Francisco, CA.

Khong, Yuen Foong. 1992. *Analogies at War.* Princeton: Princeton University Press.

King, Gary, Robert O. Keohane, and Sidney Verba. 1994. *Designing Social Inquiry: Scientific Inference in Qualitative Research.* Princeton: Princeton University Press.

Kinnvall, Catarina. 2004. Globalization and Religious Nationalism: Self, Identity and the Search for Ontological Security. *Political Psychology,* 25, no. 4: 741–67.

Kinnvall, Catarina. 2007. *The Search for Ontological Security: Globalization and Religious Nationalism in India.* London: Routledge.

Klingberg, Frank. 1952. The Historical Alternation of Moods in American Foreign Policy. *World Politics* 4, no. 2: 239–73.

Klusmeyer, Douglas. 2005. Hannah Arendt's Critical Realism: Power, Justice and Responsibility. In *Hannah Arendt and International Relations,* ed. Anthony F. Lang, Jr., and John Williams, 113–78. New York: Palgrave-MacMillan.

Kogler, Hans. 1996. *The Power of Dialogue: Critical Hermeneutics after Gadamer and Foucault.* Trans. Paul Hendrickson. Cambridge, MA: MIT Press.

Konstan, David. 1996. Friendship, Frankness and Flattery. In *Friendship, Flattery and Frankness of Speech,* ed. John Fitzgerald, 7–20. Leiden: Brill.

Krainik, Clifford. 2002. Face the Lens, Mr. President: A Gallery of Photographic Portraits of Nineteenth-Century Presidents. *White House History* 16:23–39.

Kratochwil, Friedrich. 2006. History, Action and Identity: Revisiting the "Second" Great Debate and Assessing Its Importance for Social Theory. *European Journal of International Relations* 12, no. 1: 5–29.

Kratochwil, Friedrich. 2007. Reflections on the "Critical" in Critical Theory. *Review of International Studies* 33:25–46.

Krebs, Ronald, and Patrick Jackson. 2007. Twisting Tongues and Twisting Arms: The Power of Political Rhetoric. *European Journal of International Relations* 13, no. 1: 35–66.

Kuhn, Thomas. 1962. *The Structure of Scientific Revolutions.* Chicago: University of Chicago Press.

Kundera, Milan. 1984. *The Unbearable Lightness of Being.* 68 Publishers.

Lacan, Jacques. 1977. *Ecrits: A Selection.* Trans. Alan Sheridan. New York: W. W. Norton.

Lamb, Kevin. 2005. Foucault's Aestheticism. *Diacritics* 35, no. 2 (Summer): 43–64.

Lang, Anthony F. 2002. *Agency and Ethics.* Albany: State University of New York Press.

Lang, Anthony. 2008. *Punishment, Justice, and International Relations: Ethics and Order after the Cold War.* New York: Routledge.

Layne, Christopher. 1993. The Unipolar Illusion: Why Great Powers Will Rise. *International Security* 17, no. 4: 5–51.

Leahy, Patrick. 2004. Leahy Calls for Additional Aid for Tsunami-Stricken South Asia. http://leahy.senate.gov/press/200412/122904.html (accessed 27 March 2007).

Lebow, Richard Ned. 2003. *The Tragic Vision of Politics.* Cambridge: Cambridge University Press.

Lebow, Richard Ned. 2007. The Power of Persuasion. In *Power in World Politics,* ed. Felix Berenskoetter and M. J. Williams, 120–41. New York: Routledge.

Lelyveld, Joseph. 1985. The Enduring Legacy. *New York Times Magazine,* 31 March.

Levy, Jack S. 1988. Domestic Politics and War. *Journal of Interdisciplinary History* 18, no. 4: 653–73.

Lilla, Mark. 1998. Kant's Theological-Political Revolution. *Review of Metaphysics* 52 (December): 397–434.

Linklater, Andrew. 2007. Torture and Civilisation. *International Relations* 21, no. 1: 111–18.

Little, Richard. 2007. *The Balance of Power in International Relations*. New York: Cambridge University Press.

Lomas, Peter. 2005. Anthropomorphism, Personification, and Ethics: A Reply to Alexander Wendt. *Review of International Studies* 31, no. 2: 349–55.

Löwenheim, Oded. 2003. "Do Ourselves Credit and Render a Lasting Service to Mankind": British Moral Prestige, Humanitarian Intervention and the Barbary Pirates. *International Studies Quarterly* 47, no. 1: 23–48.

Löwenheim, Oded. 2006. *Predators and Parasites*. Ann Arbor: University of Michigan Press.

Löwenheim, Oded. 2007a. The Responsibility to Responsibilize: Foreign Offices and the Issuing of Travel Warnings. *International Political Sociology* 1, no. 3: 203–21.

Löwenheim, Oded. 2007b. Examination Period: The Rise of Rating and Ranking in International Politics. Paper presented at the International Studies Association Annual Meeting, Chicago.

Löwenheim, Oded, and Gadi Heimann. 2008. Revenge in International Politics. *Security Studies* 17:685–724.

Löwenheim, Oded, and Brent J. Steele. 2010. Institutions of Violence, Great Power Authority, and the War on Terror. *International Political Science Review* 31 (forthcoming).

MacDonald, Paul. 2004. Peripheral Pulls: Great Power Expansion and Lessons for the American Empire. Paper presented at the International Studies Association Annual Meeting, Montreal, Canada.

Maoz, Zeev. 2001. Democratic Networks: Connecting National, Dyadic, and Systemic Levels-of-Analysis in the Study of Democracy and War. In *War in a Changing World*, ed. Zeev Maoz and Azar Gat. Ann Arbor: University of Michigan Press.

March, James G., and Johan P. Olsen. 1998. The Institutional Dynamics of International Political Orders. *International Organization* 52, no. 4: 943–69.

Markey, Daniel. 1999. Prestige and the Origins of War: Returning to Realism's Roots. *Security Studies* 8:126–73.

Mastanduno, Michael. 1997. Preserving the Unipolar Moment: Realist Theories and U.S. Grand Strategy after the Cold War. *International Security* 21, no. 4: 49–88.

May, Tim. 2000. A Future for Critique? Positioning, Belonging and Reflexivity. *European Journal of Social Theory* 3, no. 2: 157–73.

Mayer, Jane. 2004. The Manipulator. *New Yorker*, 7 June.

Mayer, Jane. 2006. The C.I.A.'s Travel Agent. *New Yorker*, 30 October.

Mayer, Jane. 2007. Whatever It Takes. *New Yorker*, 19 February.

Mayer, Jane. 2008. *The Dark Side: The Inside Story of How the War on Terror Turned into a War on American Ideals*. New York: Doubleday.

Mayhew, David. 2004. Review of *Realignment: The Theory That Changed the Way We Think about American Politics*, by Theodore Rosenof. *Journal of Interdisciplinary History* 35, no. 2: 321–22.

McDermott, Rose. 2004. The Feeling of Rationality: The Meaning of Neuroscientific

Advances for Political Science. *Perspectives on Politics* 2, no. 4 (December): 691–706.

McFadden, Robert D. 1971. Calley Verdict Brings Home the Anguish of War to Public. *New York Times*, 4 April.

McNay, Lois. 1994. *Foucault: A Critical Introduction.* London: Polity Press.

McSweeney, Bill. 1999. *Security, Identity, and Interests.* Cambridge: Cambridge University Press.

McWilliams, Wilson Carey. 1972. *Military Honor after My Lai.* New York: Council on Religion and International Affairs.

Mearsheimer, John J. 1990. Back to the Future: Instability in Europe after the Cold War. *International Security* 15, no. 1 (Summer): 5–56.

Mearsheimer, John J. 2001. *The Tragedy of Great Power Politics.* New York: W. W. Norton.

Mearsheimer, John. 2005. Hans Morgenthau and the Iraq War: Realism versus Neoconservatism. 18 May. http://www.opendemocracy.net/democracy-american power/morgenthau_2522.jsp (accessed 18 January 2010).

Mendelsohn, Barak. 2005. Sovereignty under Attack: The International Society Meets the al Qaeda Network. *Review of International Studies* 31, no. 1 (January): 45–68.

Mercer, Jonathan. 2005. Rationality and Psychology in International Studies. *International Organization* 59, no. 1 (January): 77–106.

Meyer, John, John Boli, George M. Thomas, and Francisco O. Ramirez. 1997. World Society and the Nation-State. *American Journal of Sociology* 103, no. 1: 144–81.

Mitrany, David. [1969] 1975. The Making of Functional Theory: A Memoir. In *The Functional Theory of Politics,* by David Mitrany. London: M. Robertson on behalf of London School of Economics and Political Science.

Mitrany, David. 1977. The Functional Approach to World Organization. In *New International Actors: The United Nations and the European Economic Community,* ed. Carol Ann Cosgrove and Kenneth J. Twitchett, 65–75. London: MacMillan.

Mitzen, Jennifer. 2006. Ontological Security in World Politics: State Identity and the Security Dilemma. *European Journal of International Relations* 12, no. 3: 341–70.

Modelski, George. 1978. Long Cycles of Global Politics and the Nation State. *Comparative Studies in Society and History* 20:214–38.

Morgenthau, Hans. [1939] 1962. Grandeur and Decadence of Spanish Civilization. In *The Decline of Democratic Politics,* 212–19. Chicago: University of Chicago Press.

Morgenthau, Hans. 1946. *Scientific Man vs. Power Politics.* Chicago: University of Chicago Press.

Morgenthau, Hans. [1947] 1962. The Escape from Power in the Western World. In *Decline of Democratic Politics,* Politics of the Twentieth Century 1. Chicago: University of Chicago Press.

Morgenthau, Hans. [1948] 2006. *Politics among Nations.* 7th ed. Boston: McGraw-Hill Higher Education.

Morgenthau, Hans. 1950. The Mainsprings of American Foreign Policy. *American Political Science Review* 44, no. 4 (December): 833–54.

Morgenthau, Hans. [1952] 1962. The Military Displacement of Politics. In *The Decline of Democratic Politics,* 328–41. Chicago: University of Chicago Press.

Morgenthau, Hans. [1959] 1962. The Commitments of a Theory of International Politics. In *The Decline of Democratic Politics,* 55–61. Chicago: University of Chicago Press.

Morgenthau, Hans. 1961. Death in the Nuclear Age, *Commentary* (September): 280–85.

Morgenthau, Hans. [1967] 1970. Modern Science and Political Power. In *Truth and Power: Essays of a Decade 1960–1970*. New York: Praeger.

Morgenthau, Hans. 1970. *Truth and Power: Essays of a Decade 1960–1970*. New York: Praeger.

Moyar, Mark. 2006. *Triumph Forsaken: The Vietnam War, 1954–1965*. New York: Cambridge University Press.

Mueller, John. 2005. Simplicity and Spook: Terrorism and the Dynamics of Threat Exaggeration. *International Studies Perspectives* 6:208–34.

Müller, Harald. 2001. International Relations as Communicative Action. In *Constructing International Relations: The Next Generation*, ed. Karin M. Fierke and Knud Erik Jorgensen, 160–78. Armonk, NY: M.E. Sharpe.

Muravchik, Joshua. 1992. *Exporting Democracy: Fulfilling America's Destiny*. Washington, DC: AEI Press.

Nash, Laura. 1978. Concepts of Existence: Greek Origins of Generational Thought. *Daedalus* 107, no. 4: 1–21.

National Public Radio. 2004. *All Things Considered*. 29 December.

Nexon, Daniel H., and Thomas Wright. 2007. What's at Stake in the American Empire Debate. *American Political Science Review* 101, no. 2: 253–71.

Niebuhr, Reinhold. 1932. *Moral Man and Immoral Society*. New York: Scribner.

Nye, Joseph. 2004. *The Paradox of American Power*. New York: Oxford University Press.

O'Driscoll, Cian. 2007. Jean Bethke Elshtain's Just War against Terror: A Tale of Two Cities. *International Relations* 21, no. 4: 485–92.

Onuf, Nicholas. 1989. *World of Our Making*. Columbia: University of South Carolina Press.

Orwell, George. 1941. The Lion and the Unicorn. http://www.orwell.ru/library/essays/lion/english/e_eye.

Orwell, George. 1962. *Burmese Days*. Time.

Osborne, Thomas. 1999. Critical Spirituality: On Ethics and Politics in the Later Foucault. In *Foucault contra Habermas*, ed. Samantha Ashenden and David Owen, 45–59. London: Sage.

Owen, David. 2006. Perfectionism, Parrhesia and Care of the Self: Cavell and Foucault on Ethics and Politics. In *The Claim to Community: Essays on Stanley Cavell and Political Philosophy*, ed. A. Norris. Stanford: Stanford University Press.

Owens, Patricia. 2005. Hannah Arendt, Violence, and the Inescapable Fact of Humanity. In *Hannah Arendt and International Relations*, ed. Anthony F. Lang, Jr., and John Williams. New York: Palgrave-MacMillan.

Owens, Patricia. 2007. Beyond Strauss, Lies, and the War in Iraq: Hannah Arendt's Critique of Neoconservativism. *European Journal of International Relations* 33, no. 2: 265–83.

Packer, George. 2002. The Liberal Quandry over Iraq. *New York Times*, 7 December.

Packer, George. 2003. Dreaming Demolition. *New York Times*, 2 March.

Packer, George. 2005. *Assassin's Gate*. New York: Farrar, Straus, and Giroux.

Packer, George. 2007. A Reporter at Large: Betrayed: The Iraqis Who Trusted America the Most. *New Yorker*, 26 March.

Peterson, John. 2004. America as a European Power: The End of Empire by Integration? *International Affairs* 80, no. 4: 613–41.

Peterson, V. I. 1992. *Gendered States: Feminist (Re)visions of International Relations Theory.* Boulder: Lynne Rienner.

Podhoretz, Norman. 1963. My Negro Problem—And Ours. *Commentary* (February): 95–103.

Prozorov, Sergei. 2007. The Ethos of Insecure Life: Reading Carl Schmitt's Existential Decisionism as a Foucauldian Ethics. In *The International Political Thought of Carl Schmitt: Terror, Liberal War and the Crisis of Global Order,* ed. Louiza Odysseos and Fabio Petito. London: Routledge.

Ralph, Jason C. 2005. International Society, the International Criminal Court and American Foreign Policy. *Review of International Studies* 31:27–44.

Reiter, Dan, and Allan C. Stam. 1998. Democracy, War Initiation, and Victory. *American Political Science Review* 92, no. 2: 377–89.

Reiter, Dan, and Allan C. Stam. 2002. *Democracies at War.* Princeton: Princeton University Press.

Rengger, Nicholas. 2002. On the Just War Tradition in the Twenty-first Century. *International Affairs* 78, no. 2: 353–63.

Reus-Smit, Christian. 1999. *The Moral Purpose of the State.* Princeton: Princeton University Press.

Ricks, Thomas. 2006. *Fiasco: The American Military Adventure in Iraq.* New York: Penguin.

Risse, Thomas. 1996. Collective Identity in a Democratic Community: The Case of NATO. In *The Culture of National Security,* ed. Peter Katzenstein, 357–99. New York: Columbia University Press.

Risse, Thomas. 1999. International Norms and Domestic Change: Arguing and Communicative Behavior in the Human Rights Arena. *Politics and Society* 27, no. 4: 529–59.

Risse, Thomas. 2000. "Let's Argue!" Communicative Action in World Politics. *International Organization* 54, no. 4: 1–40.

Risse, Thomas, Stephen Ropp, and Kathryn Sikkink. 1998. *The Power of Human Rights.* Cambridge: Cambridge University Press.

Robin, Corey. 2004. *Fear: The History of a Political Idea.* London: Oxford University Press.

Rorty, Richard. 1982. Method, Social Science, and Social Hope. In *Consequences of Pragmatism: Essays, 1972–1980.* Minneapolis: University of Minnesota Press.

Rose, Nikolas, and Peter Miller. 1992. Political Power beyond the State: Problematic of Government. *British Journal of Sociology* 43:173–205.

Roskin, Michael. 1974. From Pearl Harbor to Vietnam: Shifting Generational Paradigms and Foreign Policy. *Political Science Quarterly* 89, no. 3: 563–88.

Ross, Andrew A. G. 2006. Coming in from the Cold: Constructivism and Emotions. *European Journal of International Relations* 12, no. 2 (June): 197–222.

Roston, Aram. 2008. *The Man Who Pushed America to War: The Extraordinary Life, Adventures, and Obsessions of Ahmad Chalabi.* New York: Nation.

Ruggie, John. 1982. International Regimes, Transactions and Social Change: Embedded Liberalism in the Postwar Economic System. *International Organization* 36, no. 2: 379–415.

Russell, Greg. 1991. Science, Technology and Death in the Nuclear Age: Hans J. Morgenthau on Nuclear Ethics. *Ethics and International Affairs* 5:115–34.

Russett, Bruce. 1993. *Grasping the Democratic Peace.* Princeton: Princeton University Press.

Russett, Bruce. 2005. Bushwhacking the Democratic Peace. *International Studies Perspectives* 6, no. 4: 395–408.

Russett, Bruce, and John Oneal. 2001. *Triangulating Peace: Democracy, Interdependence and International Organizations.* New York: W. W. Norton.

Salter, Mark B. 2007. Governmentalities of an Airport: Heterotopia and Confession. *International Political Sociology* 1, no. 1: 49–66.

Sammon, Bill. 2004. U.N. Official Slams U.S. as "Stingy" over Aid. *Washington Times,* 28 December.

Sanger, Warren, and David Hoge. 2005. U.S. Vows Big Rise in Aid for Victims of Asian Disaster. *New York Times,* 1 January.

Sasley, Brent. 2010. Affective Attachments and Foreign Policy. *European Journal of International Relations* 16 (forthcoming).

Saurette, Paul. 2006. You Dissin Me? Humiliation and Post 9/11 Global Politics. *Review of International Studies* 32, no. 4: 495–522.

Schlesinger, Arthur, Jr. 1949. *The Vital Center: The Politics of Freedom.* Boston: Houghton Mifflin.

Schmitt, Carl. [1929] 2006. The Age of Neutralizations and Depoliticizations. In *The Concept of the Political,* 80–96. Chicago: University of Chicago Press.

Schmitt, Carl. 2005. *Political Theology: Four Chapters on the Concept of Sovereignty.* Chicago: University of Chicago Press.

Schroeder, Paul. 1994. Historical Reality v. Neo-realist Theory. *International Security* 19, no. 1: 108–48.

Schweller, Randall. 1994. Bandwagoning for Profit: Bringing the Revisionist State Back In. *International Security* 19, no. 1: 72–107.

Schweller, Randall. 2003. The Progressiveness of Neoclassical Realism. In *Progress in International Relations Theory: Appraising the Field,* ed. Colin Elman and Miriam Fendius Elman. Boston: MIT Press.

Schweller, Randall. 2008. *Unanswered Threats: Political Constraints on the Balance of Power.* Princeton: Princeton University Press.

Scott, James C. 1985. *Weapons of the Weak: Everyday Forms of Peasant Resistance.* New Haven: Yale University Press.

Scott, James C. 1990. *Domination and the Arts of Resistance: Hidden Transcripts.* New Haven: Yale University Press.

Semple, Robert B. 1970. Nixon Sends Combat Forces to Cambodia. *New York Times,* 1 May.

Shapcott, Richard. 2001. *Justice, Community and Dialogue in International Relations.* Cambridge: Cambridge University Press.

Shaw, Martin. 2000. *Theory of the Global State.* Cambridge: Cambridge University Press.

Shaw, Martin. 2001. The Unfinished Global Revolution: Intellectuals and the New Politics of International Relations. *Review of International Studies* 27, no. 3: 627–47.

Shiner, Larry. 2001. *The Invention of Art: A Cultural History.* Chicago: University of Chicago Press.

Simons, Jon. 1995. *Foucault and the Political.* London: Routledge.

Singer, J. David. 1961. The Levels of Analysis Problem in International Relations. In *The International System,* ed. K. Knorr and S. Verba. Princeton: Princeton University Press.

Singer, Peter. 2003. *Corporate Warriors.* Ithaca: Cornell University Press.

Sipiora, Phillip, and James S. Baumlin, eds. 2002. *Rhetoric and Kairos: Essays in History, Theory, and Praxis.* Albany: State University of New York Press.

Sjostedt, Roxanna. 2007. The Discursive Origins of a Doctrine: Norms, Identity and Securitization under Harry S. Truman and George W. Bush. *Foreign Policy Analysis* 3, no. 3: 233–54.

Slaughter, Ann-Marie. 2007. Partisans Gone Wild. *Washington Post,* 29 July.

Smith, Steve. 1999. Is the Truth out There? Eight Questions about International Order. In *International Order and the Future of World Politics,* ed. T. V. Paul and John Hall, 99–119. New York: Cambridge University Press.

Smith, Steve. 2002. The United States and the Discipline of International Relations: "Hegemonic Country, Hegemonic Discipline." *International Studies Review* 4, no. 2: 67–86.

Smith, Tony. 2007. *A Pact with the Devil.* New York: Routledge.

Sorley, Lewis. 1999. *A Better War: The Unexamined Victories and Final Tragedy of America's Final Years in Vietnam.* Orlando, FL: Harcourt.

Spruyt, Hendrik. 2005. *Ending Empire: Contested Sovereignty and Territorial Partition.* Ithaca: Cornell University Press.

Spykman, Nicholas. 1942. *America's Strategy in World Politics: The United States and the Balance of Power.* New York: Harcourt, Brace.

Steele, Brent J. 2005. Ontological Security and the Power of Self-Identity: British Neutrality and the American Civil War. *Review of International Studies,* 31, no. 3: 519–40.

Steele, Brent J. 2007a. Liberal-Idealism: A Constructivist Critique. *International Studies Review* 9, no. 1: 23–52.

Steele, Brent J. 2007b. "Eavesdropping on Honored Ghosts": From Classical to Reflexive Realism. *Journal of International Relations and Development* 10, no. 3: 272–300.

Steele, Brent J. 2007c. Making Words Matter: The Asian Tsunami, Darfur, and "Reflexive Discourse" in International Politics. *International Studies Quarterly* 51, no. 4 (December): 901–25.

Steele, Brent J. 2008a. *Ontological Security in International Relations.* London: Routledge.

Steele, Brent J. 2008b. "Ideals That Were Never Really in Our Possession": Torture, Honor, and U.S. Identity. *International Relations* 22, no. 2: 243–61.

Steele, Brent J., and Jack Amoureux. 2005. NGOs and Monitoring Genocide: The Benefits and Limits of a Human Rights Panopticon. *Millennium: Journal of International Studies* 34, no. 2: 403–32.

Stephney, M. J. 2008. Gen X: the Ignored Generation? *Time,* 16 April. http://www.time.com/time/arts/article/0,8599,1731528,00.html.

Sterling-Folker, Jennifer, and Rosemary Shinko. 2005. Discourses of Power: Traversing the Realist-Postmodern Divide. *Millennium: Journal of International Studies* 33, no. 3: 637–64.

Strauss, W., and N. Howe. 1991. *Generations: The History of America's Future, 1584 to 2069.* New York: Morrow.

Sucharov, Mira. 2006. *The International Self: Psychoanalysis and the Search for Israeli-Palestinian Peace.* Albany: SUNY Press.

Suskind, Ronald. 2006. *The One Percent Doctrine.* New York: Simon and Schuster.

Tammen, Ronald L., et al. 2000. *Power Transitions: Strategies for the 21st Century.* New York: Chatham House.

Thompson, Kenrick S., Alfred C. Clarke, and Simon Dinitz. 1974. Reactions to My-

Lai: A Visual-Verbal Comparison. *Sociology and Social Research* 58, no. 2 (January): 122–29.

Tickner, J. Ann, and Laura Sjoberg. 2006. Feminism. In *International Relations Theories: Discipline and Diversity*, ed. Tim Dunne, Milja Kurki, and Steve Smith. Oxford: Oxford University Press.

Tjalve, Vibeke Schou. 2008. *Realist Strategies of Republican Peace: Neibuhr, Morgenthau and the Politics of Patriotic Dissent.* New York: Palgrave-MacMillan.

de Tocqueville, Alexis. 1969. *Democracy in America.* Ed. J. P. Mayer. Trans. George Lawrence. Garden City, NY: Doubleday.

Toynbee, A. J. 1954. *A Study of History.* Vol. 9. Oxford University Press.

Traub, James. 2006. *The Best Intentions: Kofi Annan and the UN in the Era of American World Power.* New York: Farrar, Strauss and Giroux.

Tully, James. 1999. To Think and Act Differently: Foucault and Four Reciprocal Objections to Habermas' Theory. In *Foucault contra Habermas*, ed. Samantha Ashenden and David Owen, 90–142. London: Sage.

Uphoff, Norman. 1989. Distinguishing Power, Authority and Legitimacy: Taking Max Weber at His Word. *Polity* 22, no. 2: 295–322.

United Nations Office of the Special Envoy for Tsunami Relief. 2006. 7 May. http://www.tsunamispecialenvoy.org/country/humantoll.asp.

United States Department of State. 2004a. Secretary of State Powell's Testimony before the Senate Foreign Relations Committee. 9 September. http://www.state.gov/secretary/former/powell/remarks/36042.htm.

United States Department of State. 2004b. Daily press briefing. 28 December. http://www.state.gov/r/pa/prs/dpb/2004/40083.htm.

United States Department of State. 2004c. Secretary of State Powell's Interview on ABC's Good Morning America with Christopher Cuomo. http://www.state.gov/secretary/former/powell/remarks/40071.htm (accessed 15 January 2008).

Van Es, Hubert. 2005. Thirty Years at 300 Millimeters. *New York Times,* 29 April 2005. http://www.nytimes.com/2005/04/29/opinion/29van_es.html.

Vertzberger, Yaacov. 1990. *The World in Their Minds: Information Processing, Cognition, and Perception in Foreign Policy Decisionmaking.* Stanford: Stanford University Press.

Vertzberger, Yacov. 2005. The Practice and Power of Collective Memory. Review of *Trauma and the Memory of Politics,* by Jenny Edkins. *International Studies Review* 7, no. 1: 117–20.

Wæver, Ole. 1993. Societal Security: The Concept. In *Identity, Migration and the New Security Dilemma in Europe*, ed. Ole Wæver, Martin Kelstrup, and Pierre Lemaitre. London: Pinter.

Wæver, Ole. 1995. Securitization and Desecuritization. In *On Security*, ed. Ronnie D. Lipschutz, 46–85. New York: Columbia University Press.

Wæver, Ole. 1998. The Sociology of a Not So International Discipline: American and European Developments in International Relations. *International Organzation* 2:687–727.

Walker, Stephen G. 1978. International Restraints and Foreign Policy Choices: British Behavior in Military Situations: 1931–1941. *International Interactions* 5:39–65.

Walt, Stephen. 1987. *The Origins of Alliances.* Ithaca: Cornell University Press.

Walt, Stephen. 1991. The Renaissance of Security Studies. *International Studies Quarterly* 35:211–39.

Walt, Stephen. 2006. Letter to the editor. *Commentary* (January): 3–4.

Waltz, Kenneth. 1959. *Man, State and War.* New York: Columbia University Press.

Waltz, Kenneth. 1979. *Theory of International Politics*. Reading, MA: Addison-Wesley.

Waltz, Kenneth. 1993. The Emerging Structure of International Politics. *International Security* 18, no. 2: 44–79.

Watson, Adam. 1992. *The Evolution of International Society*. London: Routledge.

Weaver, Catherine. 2008. *The Hypocrisy Trap*. Princeton: Princeton University Press.

Weaver, Jace. 1995. Original Simplicities and Present Complexities: Reinhold Niebuhr, Ethnocentrism, and the Myth of American Exceptionalism. *Journal of the American Academy of Religion* 63, no. 2: 231–47.

Weber, Max. [1922] 1949. *The Theory of Social and Economic Organization*. Trans. A. R. Henderson. Ed. Talcott Parsons. New York: Oxford University Press.

Weber, Max. [1956] 1976. *Economy and Society*. Berkeley: University of California Press.

Weldes, Jutta. 1999. The Cultural Production of Crises: U.S. Identity and Missiles in Cuba. In *Cultures of Insecurity,* ed. Jutta Weldes, Mark Laffey, Hugh Gusterson, and Raymond Duvall. Minneapolis: University of Minnesota Press.

Wendt, Alexander. 2003. Why a World State Is Inevitable. *European Journal of International Relations* 9, no. 4: 491–542.

Wendt, Alexander. 2005. How Not to Argue against State Personhood: A Reply to Lomas. *Review of International Studies* 31, no. 2: 357–60.

West, Bing. 2005. *No True Glory: A Frontline Account of the Battle for Fallujah*. New York: Bantam Books.

Westbrook, Robert B. 1991. *John Dewey and American Democracy*. Ithaca: Cornell University Press.

Wheeler, Nicholas. 2000. *Saving Strangers*. Oxford: Oxford University Press.

Wight, Martin. 1991. *International Theory: The Three Traditions*. Leicester: Leicester University Press/Royal Institute of International Affairs.

Williams, Michael C. 2003. Words, Images and Enemies. *International Studies Quarterly* 47, no. 4: 511–31.

Williams, Michael C. 2004. *The Realist Tradition and the Limits of International Relations*. Cambridge: Cambridge University Press.

Williams, Michael C. 2005. What Is the National Interest? The Neoconservative Challenge in IR Theory. *European Journal of International Relations* 11, no. 3 (September): 303–37.

Williams, Michael C., ed. 2007. *Realism Reconsidered: The Legacy of Hans Morgenthau in International Relations*. Oxford: Oxford University Press.

Williams, Michael J. 2007. Theory Meets Practice: Facets of Power in the "War on Terror." In *Power in World Politics,* ed. Felix Berenskoetter and M. J. Williams, 265–76. New York: Routledge.

Wilmer, Franke. 2003. "Ce n'est pas une Guerre/This Is Not a War": The International Language and Practice of Political Violence. In *Language, Agency, and Politics in a Constructed World,* ed. Francois Debrix, 220–46. New York: M. E. Sharpe.

Wilson, John K. 2005. Academic Freedom in America after 9/11. *Thought and Action* (Fall): 119–35.

Winograd, Morley, and Michael D. Hais. 2008. *Millennial Makeover.* New Brunswick: Rutgers University Press.

Wohlforth, William. 1999. The Stability of the Unipolar World. *International Security* 24, no. 1: 5–41.

Wolf, Naomi. 1991. *The Beauty Myth: How Images of Beauty are Used against Women*. New York: W. Morrow.

Wolfers, Arnold. 1952. National Security as an Ambiguous Symbol. *Political Science Quarterly* 67, no. 4: 481–502.

Wolfers, Arnold. 1962. *Discord and Collaboration*. Baltimore: Johns Hopkins University Press.

Wolin, Richard. 1992. Carl Schmitt, the Conservative Revolutionary: Habitus and the Aesthetics of Horror. *Political Theory* 20, no. 3: 424–47.

Youde, Jeremy. 2005. The Development of a Counter-Epistemic Community: AIDS, South Africa, and International Regimes. *International Relations* 19, no. 4: 421–39.

Zarakol, Ayse. 2010. Ontological (In)security and State Denial of Past Crimes: Turkey and Japan. *International Relations* 24, no. 1: 3–23.

Zehfuss, Maja. 2007. *Wounds of Memory*. Cambridge: Cambridge University Press.

Žižek, Slavoj. 1989. *The Sublime Object of Ideology*. New York: Verso.

Index

Neoconservatism, 10, 62n, 67, 191
  versus Arendtian analysis, 61n43
  and Fallujah, 149–51
  Podhoretz's, 54–58, 145–46
  and torture, 158–59
  and 2004 Asian Tsunami, 95–97
  and Vietnam, 143–48
  vitalism of, 54–58, 96, 143–47,
    149–51
Niebuhr, Reinhold, 4, 8, 11, 15, 29,
  141, 179, 184–85, 202, 205,
  209n11

Orwell, George, 15, 193–94, 211
Other, 46, 52, 103n1, 106, 149, 153,
  158, 166–67
  role in power analyses, 183–85

Paradigms, 63–68, 109, 195
Parrhesia, 10, 22, chapter 3 passim
  academic-intellectual, 22–23, 65, 92n,
    119–32
  Cynic, 22–23, 112–19
  and flattery, 105–7
  as force, 104–5
  and friendship, 101, 103, 106, 113,
    130–31
  and hypocrisy, 108, 111–12
  and Rod Ridenhour, 137, 139–40
  tact and, 109–12
Podhoretz, Norman, 54–58, 96, 144–45
Polity (and democratic "ranking"), 201
Powell, Colin, 94, 98–99
Power, 1–2
  aesthetics and, 8–9, 25, chapter 1
    passim, 38–54
  ambiguity of, 134–35, 148
  definitions, 14–17, 25
  imaginative, 2, 4, 30–36, 37–38, 100,
    108, 118, 135, 142, 204, 206
  operating in dark, 134–35, 142–43,
    158–59, 161, 183, 199–200
  psychological, 2, 4, 28–30, 37–38, 64,
    77, 100, 108, 118, 135, 204, 206
  rhythmic, 2, 4, 36–37, 37–38, 40–41,
    77, 100, 108, 109–10, 118, 135,
    204, 206
  Self, 7–8
  ugly, 142, 145, 162, 193–94
  See also Transcendental analysis of
    power; Transgressional analysis of

power; Transitional analysis of
power

Reflexive discourse, chapter 2 passim
  and Asian Tsunami, 73–75, 93–99
  versus communicative discourse,
    84–90
  limits of, 199–200
  and torture, 154–55
Restraint (and power), 29, 175, 203–7
Ricks, Thomas, 34, 80, 150–51, 156,
  198
Ridenhour, Rod, 137, 139–40
Roskin, Michael, 65–66, 68–69, 110n5,
  135–36

Schmitt, Carl, 55, 58–62, 64n47, 69
Security
  aesthetics of (*see* insecurity, aesthetics
    and)
  ambiguous symbol, 13–14
  defined, 13–14
  ontological, 15, 36, 37, 74, 78, 81,
    99–100, 109, 179
  physical, 14, 51, 162
Self
  flattery's inflation of, 74, 78–80
  humans, 2–3, 4
  nation-state, 3–4
  reflexive discourse and, 76–79
Self-interrogative imaging, chapter 4
  passim, 203–6
  defined, 23, 31–32, 133
  and generations, 134–35
Stoics, 135n, 204–7

Tjalve, Vibeke Schou, 5, 127n, 202, 208
Tocqueville, Alexis de, 62, 70, 112
Torture, 100, 124, 126, 136, 152–62,
  186, 198–201
Transcendental analysis of power,
  86–88, 169–73, 173–76, 190–91
Transgressional analysis of power, chapter 5 passim
  method, 185–86
  roles of Self and Other in, 183–85
  spatiality of, 178–82
  temporality of, 176–78
Transitional analysis of power, 166–69,
  173–76, 191
Trauma, 9, 27, 37, 63, 69–70, 118, 146